Charlie Fink's Metaverse

An AR Enabled Guide to AR & VR

By Charlie Fink

With
Matt Miesnieks, Stephanie Llamas,
Michael Eichenseer, Annika Steiber,
Mark Billinghurst, Ori Inbar,
Peter Wilkins, Dirk Schart, Samuel
Steinberger, Walter Greenleaf,
Tim Kashani, Tony Parisi, Kyle Menick,
Samantha G. Wolfe, Tim Merel,
Tamiko Theil & Zenka

Edited by Robert Fine

Foreword by Philip Rosedale,
Founder of Second Life

An Augmented Reality Book

by Living Popups

"Charlie Fink's Metaverse -

An AR Enabled Guide to VR & AR"

By Charlie Fink Foreword by Philip Rosedale

Charlie Fink (2018).

Charlie Fink's Metaverse -

An AR Enabled Guide to VR & AR. Technology.

1. Fink, Charlie

ISBN 978-1-64007-979-3

LCCN 2017964169

http://metaverse.media

Manufactured in the United States of America

First Edition

Published and distributed in the United States by

Cool Blue Media

NOTICE!

This book is out of date.

It was out of date the day it was written. That's one of the things that makes AR, VR one of the greatest stories of our time and also explains why no one else has written it.

Go to Finkmetaverse.com and enter your email for free updates of this book.

HOW TO USE THIS BOOK

The augmented book is growing in popularity, but by no means ubiquitous. Though many children's books are now beginning to embrace it. Our children are born today into a world of screens, and it seems intuitive to them that there exists a digital layer of meaning attached to complex objects like cars, and simple ones, like this book.

Download the Fink Metaverse app before you read this book. Once in the app, hold your smartphone or tablet over the page with an icon like this:

SCAN THIS ←

The square icon will activate an animated scene. Circular icons activate videos, each related to the content of the chapter.

ACKNOWLEDGEMENTS

First and foremost, I would like to thank my editor and publisher Robert Fine, without whose encouragement I would not have written this book, and to Cheryl Bayer and Jamie Dixon of Living Popups, who saw the opportunity to bring this book to life with mobile AR and animation.

Big shout out to Harriet Spear for the incredible book design; to Ontario Britton, for creating the Fink Metaverse app; and our editorial assistants (and contributors!), Samuel Steinberger, Michael Eichenseer and Sean Springle (graphics).

Special thanks to Humaneyes Technologies and Jim Malcom, who have been the most gracious sponsors an author could wish for, and to our Patron level Kickstarter backers: Ken Ehrhart, Frank Fink, Martin Tarr, Keith Boesky, Jeff Drew, Guido Van Nispen, John Arlen Parker, Bob Sopko, Chris James, Astrid Kahmke, Dominick Stirpe, John Buzzell, Stephanie Greenall, John Nguyen, Lisa Winter and John Westra. Thank you.

I must single out my friends and former colleagues, Jordan Weisman and Tim Disney, with whom I took my first journey into this incredible world in 1992.

I'd especially like to thank my contributors for their invaluable help and generosity. This was a community effort.

My writing has been encouraged, supported, published and promoted by an incredible community which includes Robert Fine (VR Voice), Lewis D'Vorkin (LA Times), Bob Drogin (LA Times), Helen Popkin (Forbes), Helen Situ (Virtual Reality Pop), Addy Khan (Cinematic VR), Malia Probst, Jesse Damiani and Jonathan Nafarrete (VR Scout), Keram Malicki-Sanchez and Jessy Blaze (VR Toronto), Ernest Cline (author, Ready Player One), Rob Tercek (author, Vaporized), John Werner (Meta), Jody Arlington (SXSW), Jim Chabin (The VR Society), Kent Bye (Voices of VR), Laura Mingail (Entertainment One), Robert Scoble and Shel Isreal (authors, Fourth Transformation), Dean Takahashi (Venture Beat), Irena Cronin (Transformation Group), Patrick Ambron, Ryan Erskine, Carlos Iniguez (Brand Yourself), Aleks Kang (Gannett) and Gary Vaynerchuk (author, founder VaynerMedia), who asked me in early 2015, "if you don't have a blog, how are you going to write a book?"

I'd like to thank the many executives, developers, entrepreneurs, thought leaders, and creators who have taken the time to speak with me over the past two years. I could not possibly name you all here, but the long list, in no particular order, includes Pattie Maes (MIT), Cathy Hackl, Rikard Steiber and Alvin Graylin (HTC), Yelena Rachitsky (Oculus), Philip Rosedale (High Fidelity), Michael Hoffman, Mike Pell and Greg Sullivan (Microsoft), Keith Boesky (ODG), Tipatat Chennavasin (The VR Fund), Amitt Mahajan (Presence Capital), Robert Sopko (Case Western Reserve), Kevin Wall, Walter Parkes and Bruce Vaughn (Dreamscape Immersive), Tim Ruse and Mike Cooney (Zero Latency), Brent Bushnell & Eric Gradman (Two Bit Circus), Ben Taft (Mira), Jeffrey Powers (Occipital), Pete Forde (itsme), Simon Che de Boer (realityvirtual.co), Shauna Heller (AiSolve), Tim Merel (Digi-Capital), Rémi Rousseau (Mimesys), Rene Schulte (Valorem), Justin Barad (Osso VR), Skip Rizzo (USC), Fred Rose, Mark Turner and Marcie Jastrow (Technicolor), Eugene Chung (Penrose), Eric Darnell and Maueen Fan (Baobab), Sam Barberie (SuperData), Jeremy Kenisky (Merge VR), Jonny Cosgrove (meetingRoom.io), Steve McNelley (DVE Telepresence), Rachel Sibley (Leap Motion), Patty Rangel (Arena Stage), Eric Romo (AltSpaceVR), Ebbe Altberg, Bjorn Laurin and Jason Gholston (Linden Lab), Michael Counts (The Ride, Paradiso), Loren Hammonds (Tribeca Interactive), Shauna Heller (AiSolve) Steve Raymond (8i), Ken Perlin, Brian Hui and Ido Lechner (NYU), Dan Burgar (Archiact), Jake Rubin and Joe Michaels (HaptX), James George and Alexander Porter (DepthKit), Yasmine Elayat (Scatter), Nathie Dejong and Sdjin "Rowdy Guy" Servaes (VR game critics), Kobi Snir, Sanchar "Vice" Weiss (AR app devlopers), Chip Sineni (Trixie Studios), Mike Levine (Happy Giant), and Bryan Biniak, who encouraged me to turn my focus back to VR.

Tech writers know the PR and communications teams are the unsung heroes of the industry. To name just a few of them: Beth Handoll (ODG), Patrick Seybold (HTC), Katya Hutnik (Oculus), Elena Caldwell (Microsoft), Erin Davern (Acer), Scott Rupp (Virtra), Ethan Rasiel and Jennifer Guerra (AWE, Kopin, Re'flekt), Kayla Holmes (IBM), Gerry Gottheil (AltSpaceVR), and Scott Accord (VR Society).

Special thanks to my friend and mentor, producer Alex Rose, who gave a hustling young guy a break in 1985, and the enlightened people who hired me, donated to my causes, invested in my companies and my shows, or simply shared their talent with me. I am filled with gratitude whenever I think of: Peter Schneider, Tim Disney, Danny Krifcher, Dan & Margaret Loeb, Kevin Wall, Rob Jennings, Joseph Mandelbaum, Jan Brandt, Mark Walsh, Steve Salzinger, Frank Weil, Ted Leonsis, Bill Haney, Stephen Prince, L. Stuart Vance, Mike Tucker, Catherine Adler, Alan Goodman, Kevin Wall, Paul Prokop, Martin Grant, Scollay Petry, Paul DeBenedictis, Michael Waxman-Lenz, John

Backus, Steve Schaffer, Abby Disney, Lee Seymour, Rob & Stefanie Sigal, Joy & Gary Every-Wortman, and Bill & Dana Schreiner, Steve Willensky, Jim Bankoff, Jon & Maria Jackson, Andy Halliday, Ted Gavin, Ted & Laura May.

Finally, and foremost, I am blessed with the love of three wonderful children and a wife who has patiently lived through at least four re-inventions. So far.

TABLE OF CONTENTS

Richard Feynman famously said "there's plenty of room at the bottom". What he meant was that atoms were incredibly small - much, much, much smaller than our eyes can see. And what this meant was that there could be an enormous amount of complicated stuff built 'down there'. And this also means that inside our computers, there is plenty enough room to build completely new worlds! As our machines get faster, they are going to get plenty fast enough to fool our senses and make us absolutely certain that we are in places as real as our waking lives. Virtual worlds with 'atoms' a fraction of a millimeter in size are going to be easy for modern computers to build, store, and simulate. There is no magic elixir that makes the real world 'real', and the virtual world 'virtual' - these are just convenient words we use to separate the old from the new. The fact is that we are at the edge of an era during which we will be able to completely re-invent, contract, or expand by a billion-fold the spaces that surround us and separate us from one another:

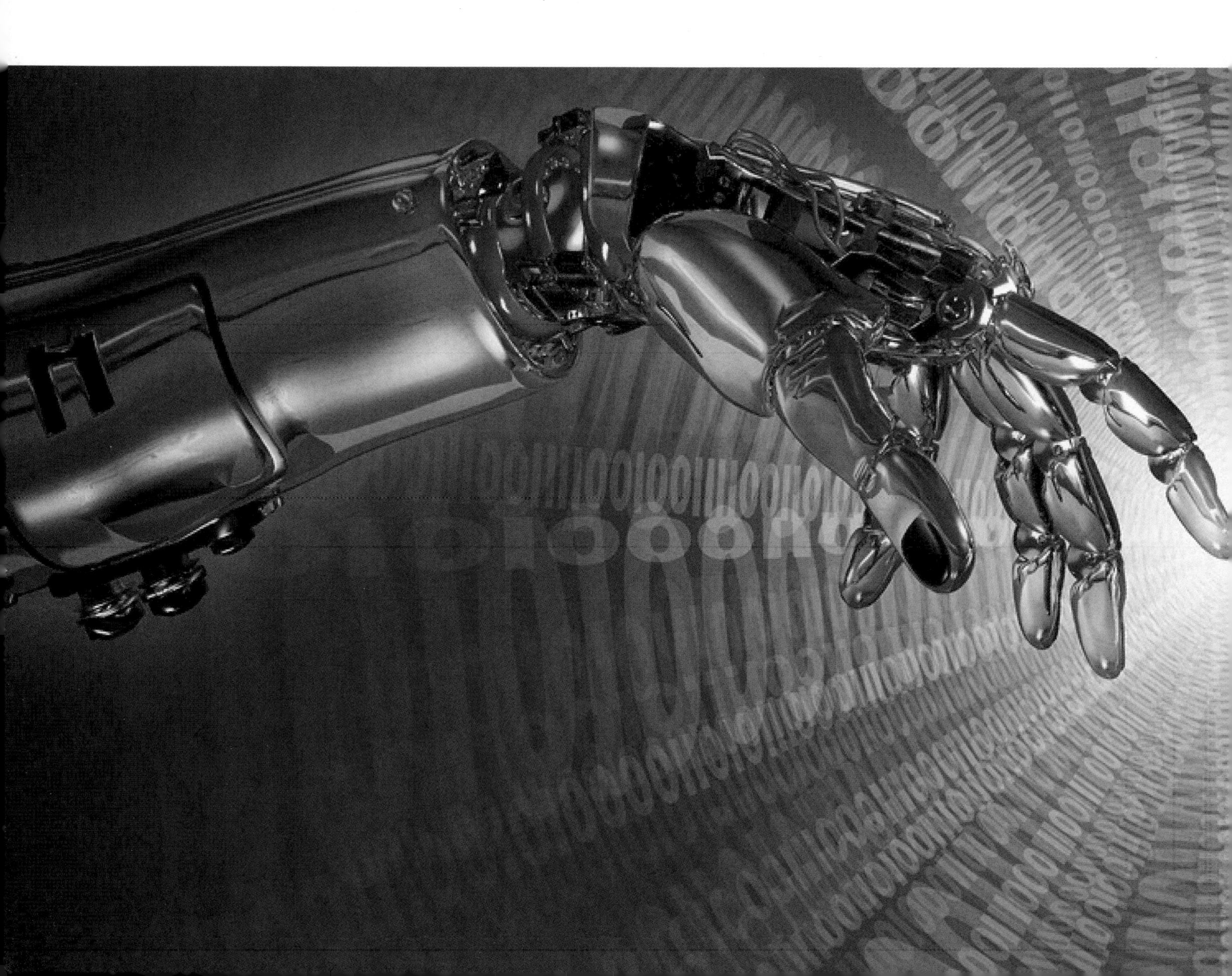

We will...

...travel across the world in the blink of an eye for a face to face meeting that begins and ends with a warm handshake and a look in the eye.

...join a group tour of a space station, led by the astronauts who live there. Or travel to the caves at Lascaux, or the bottom of the Marianas trench.

...be able to turn back time and take a few hours on the weekend to live in a perfect re-creation of a western city populated by live actors.

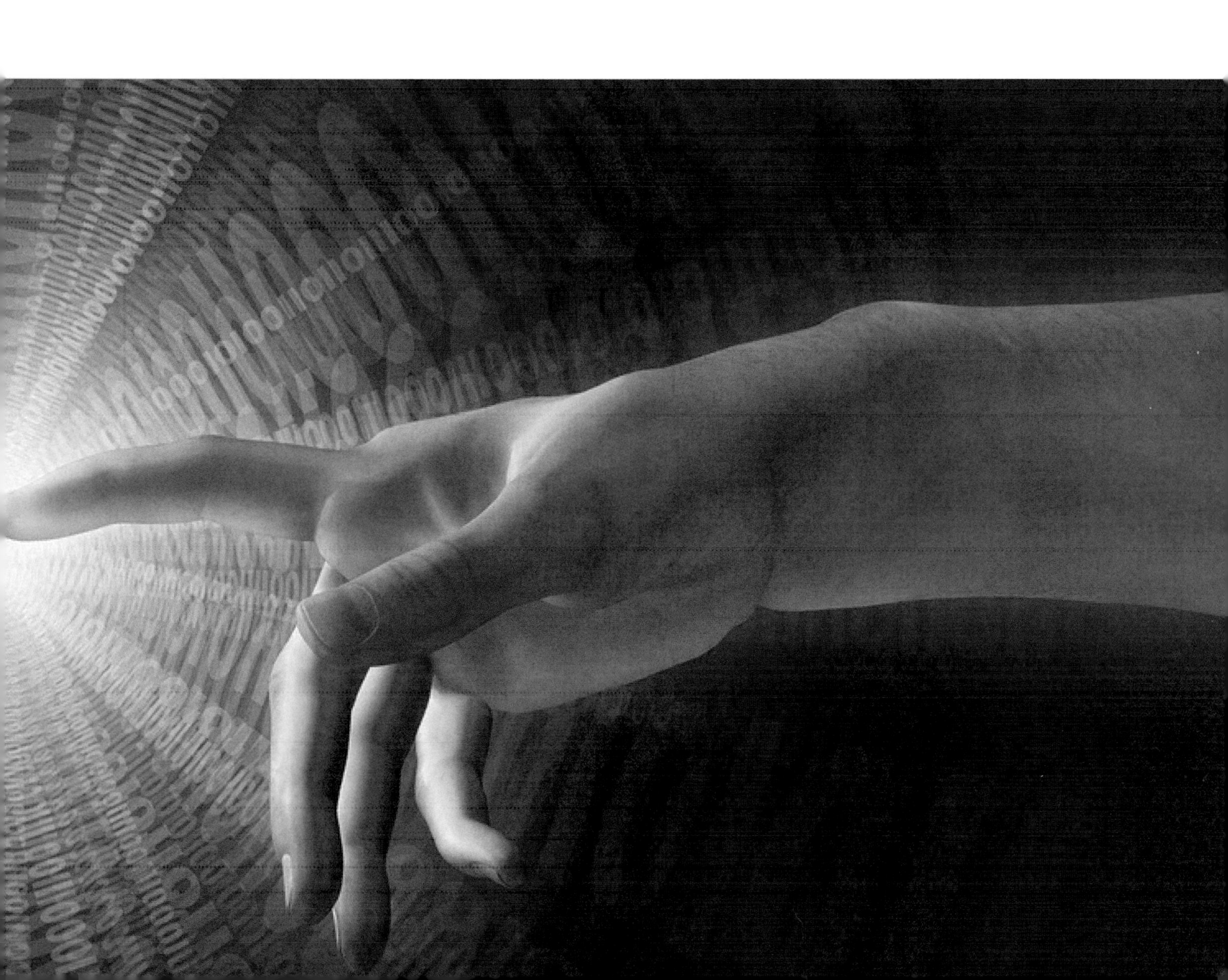

Rather than travel into outer space, we will don our headsets and travel to 'inner' space to visit planets evolved by teraflops of processing power and containing strange new organisms and ecosystems never seen before by human eyes, because they will be so much bigger than earth that we cannot possibly hope to explore but a tiny fraction of their surfaces in our lifetimes.

I've worked on VR (virtual reality) and virtual worlds myself for more than 20 years, starting as Charlie did in the early nineties with prototypes of full-body interfaces, and then starting and leading the company that in 2003 launched Second Life, in many ways the first generation of both VR and cryptocurrency.

The economy of Second Life is close to a billion dollars spread over a million people. Scale that up to the billion people Mark Zuckerberg says he wants to bring into VR and you have a trillion dollar economy and the 15th largest country in the world. That economic ride is going to be exciting and very disruptive.

VR and AR (augmented reality) will accelerate to a global stage issues of privacy and identity. Is it OK to have one company know everywhere you go, record everywhere you look and every word you say, and even be able to identify you from the movement of your body?

Much of the excitement building around 'blockchain' - public, decentralized databases for anything - will see immediate application in VR, as we suddenly will want an accounting of our digital assets and digital identity that is solely under own our control.

It is rare to have the fortune to work on something much bigger than yourself, and VR, like the Internet, is one of those things. Alongside and accompanied by AI (artificial intelligence), it will be the biggest thing, like it or not, we as humans have ever created. It is exciting to try to navigate and forecast the amazing changes that will come with these developments. Usually (and I suppose here I am biased) the people who get a really clear view of the future do so because they are single-mindedly obsessed with making some specific project come to fruition, and need to see the roadblocks ahead. Less commonly, as is the case with Charlie, they combine sharp intelligence with curiosity and the vantage point of many different projects and a journalist's eye for asking big questions and pulling it all together. His experience and perspective makes me think of Blade Runner's Roy Batty, saying "I've seen things you people wouldn't believe."

Charlie says another thing: That the killer app is other people, and this observation has carried through all my own work and rings truer than ever. And it's true for authors as well. A complex subject is only taken on with a great writer and thinker at your side. Read on!

Philip Rosedale is the CEO and co-founder of High Fidelity, Inc., a company devoted to exploring the future of next-generation shared virtual reality. Prior to High Fidelity, Rosedale founded the company Linden Lab which created the virtual civilization Second Life.

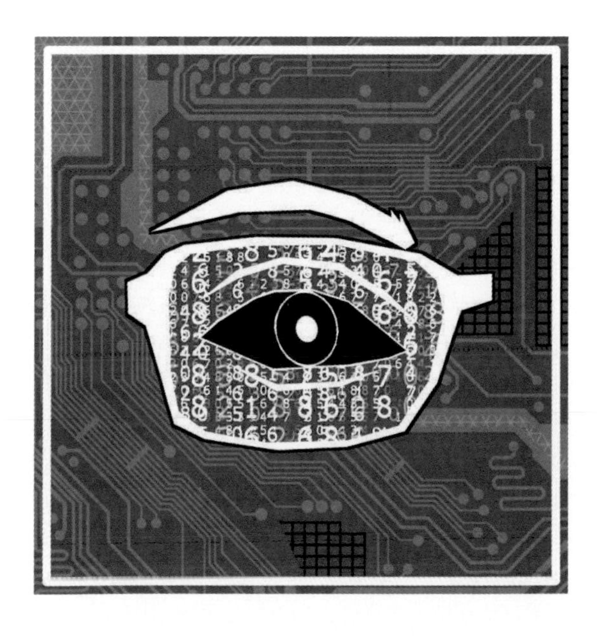

The story of the development and commercialization of VR and AR is one of the greatest business stories of our time, featuring larger than life characters, the biggest companies in the world, and huge international competitors. Hundreds of billions of dollars and the future of mankind, basically, are at stake. No one knows for sure how it's going to turn out.

Virtual, augmented and mixed reality represent a disruption in both desktop and mobile computing potentially as transformative as the smartphone, which has changed life as we know it. The march toward head mounted displays will be a shuffle at first, but it is inevitable that within ten years many of us will be wearing interoperable (AR/VR) head mounted displays at work and frequently for leisure. They will be lightweight and fashionable. As consumer computing shifted from laptops to smartphones, so will telecommunications switch from phones to headsets, but not as fast or in the way we would like them to.

My career has entered its third act. Despite our culture's vampiric obsession with youth, there are benefits to being one of the old guys. My unique career has taken me from movies to VR, from online services to social media, from local theater to Broadway, and then back to tech again. Every disruption in computing has impacted me personally and professionally. I have worked for large media conglomerates like Disney in the 80s with seemingly endless resources, to fast growing tech companies like AOL in the 90s, and tiny startups bootstrapped with a few thousand dollars, one of which I sold for forty million dollars just twelve months after I had the idea. I will skip my failures, though they provide more learning than success.

In my life I've won and I've lost. I've made history, and I've wasted time. My experiences have led me to a few basic beliefs which form the filter through which I view this story of technology, new media and culture: (1) Technology succeeds when it makes what we are doing better, faster and cheaper; (2) The killer app is other people; (3) We always overestimate the short term and underestimate the long term.

I started my career as a junior executive charged with the development of Disney animated features in 1985, working on *Little Mermaid, Beauty and the Beast, Aladdin and The Lion King,* which was based on my original idea, Bambi in Africa. I probably learned more in this job than any other. Its main lesson is hardly surprising: When you have a brand, resources, talented people and distribution, you have a huge margin for error. We spent a year and millions of dollars on a false start of our original animated production of Beauty and the Beast in 1988. It was a total whiff. It took guts for us to own that. We needed a do-over, and we got one. It was a drop in the bucket compared to the billions of dollars generated by the one most lucrative, evergreen intellectual properties ever developed by the studio. This is why so many people in the industry believe the winners, Apple,

Technology succeeds when it makes what we are already doing better, faster and cheaper.

The killer app is other people.

We always overestimate the short term and underestimate the long term.

Google, Facebook and Microsoft, have already won. They may strike out, but there will be another at bat, and another. Startups have no such luxury.

From 1992 to 1995, I was COO of Virtual World, a location-based virtual reality (LBVR) company founded by world famous role-playing game designer Jordan Weisman (Battletech, Shadowrun), and backed by Tim Disney and the Simon Property Group among others. We used modified flight simulators to network users (we called them pilots) into vehicle based simulations set in pixel friendly environments like a mining facility on Mars. The system was an ingenious kluge of Apple computers and an advanced graphics card from Silicon Graphics. Although the VRcade system we developed is still in use today, its cutting edge mid-90s technology was soon surpassed by Internet connected home computers. We even competed with ourselves, releasing a networked PC version of our premiere Battletech game through Activision.

After Virtual World, I became SVP at fast-growing America Online, which had just overtaken its rival Compuserve as the leading provider of online services. AOL charged a dollar an hour for the first ten hours, and then flipped to a $2.95 hourly rate. As Chief Creative Officer of AOL Studios, I supervised the development of original content for the formative Internet. In 1995, there were no web sites, no

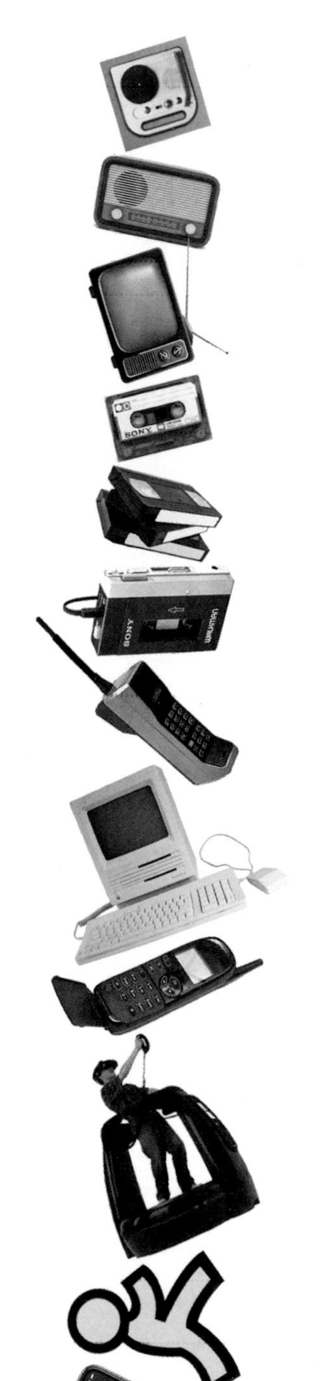

eBay, and no search engines. AOL, Prodigy and other companies had to both create and incentivize others to create content for their platforms, which is exactly what we see Sony, HTC and Oculus doing for virtual reality today.

Of all the content we developed at AOL Studios, the biggest hit was also the cheapest and easiest: Love@AOL, our free personals network, which was later sold to Match.com. The site was visited by more than half of the adults on the service proving, as did chat rooms and instant messages, that people were the killer app of AOL. It was a short hop, skip and a jump from Love@AOL personals to Facebook. It just took someone ten years to see it.

When I started to re-evaluate the field of VR several years ago, I didn't have a good place to start. There are only a handful of feature articles about virtual reality, usually the gushy "It's going to change the world!" variety. There are many more technical articles about specific technologies, problems, tools and applications. Mostly, there are a lot of short, rewritten press releases with clever click bait headlines that seem to alternate between VR has and VR hasn't arrived. For these reasons, there is nothing comprehensive on AR or VR for the general interest reader that paints a big picture of what is going on.

Voltaire said "To get rich, determine where gold will be discovered in the future, and be there when it arrives". For example, in 1990, the only people making money in personal computers were selling them to businesses. By 2000, many of those people were rich. They weren't particularly gifted, but they made one big choice, to be where money will arrive in the future. Many who missed the first wave caught the second, or the third. This is another one.

While mobile AR is emerging and gaining traction, VR sales are down this year. The medium is not going to hit a tipping point with consumers until its killer apps become clear. The Internet, in my view, has four killer apps that make it indispensable: email and messaging, ecommerce, search and social media. There are as yet no equivalent applications for VR and AR, though there are early signs of what those could be.

I asked veteran technology writers why no one has yet undertaken a definitive overview of VR and AR. The uniform response is that the book will be out of date the day it is published. That's how quickly things are moving. To address this, we intend to update the electronic version of this book frequently, adding links to relevant topics and revising or replacing chapters as needed at Finkmetaverse.com.

I am greatly indebted to my contributors for taking the time to share their deep knowledge and incredible insights to this book. I've learned a lot from them. Thought leadership is incredibly important at the birth of a new medium, and for the same reason there are a lot of conferences: No one knows what to think other than "This is really big!"

No matter which edition of this book you read, this phase in the development of personal computing will remain one of the great stories of our time, as impactful as the Internet and mobile computing.

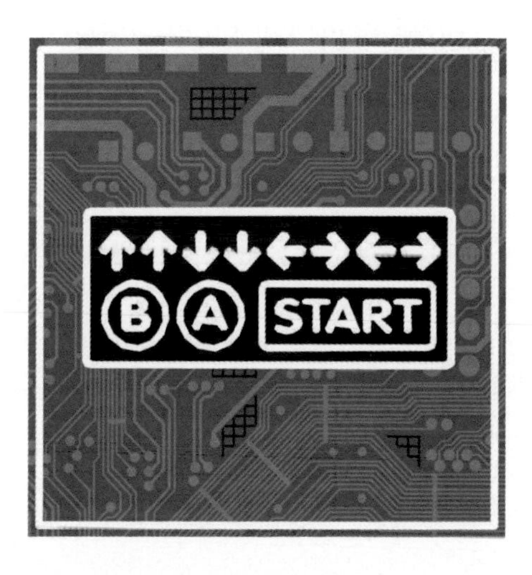

We Don't Know What We Are **Talking About**

The worst thing you can do with new technology is confuse consumers. America Online (AOL) was game changing for the Internet because all you had to do was use their tools and you could easily navigate a complex web of information. You didn't have to be bothered with understanding protocols or finding ways to quickly communicate with others. It just worked. It gave you ways to connect, communicate and explore without needing to understand its inner workings.

In the 90s, AOL used easily understandable terminology like "email", "search", and "instant messaging" that made it simple and intuitive for inexperienced users to instantly understand the functionality and value proposition. Years later, terms like Gmail, App Store and "Liking" posts are all terminology that clearly tell the user what it is. Even my 5 year old nephew and my 79 year old grandmother understand what those terms mean and how to use them.

But we are now faced with technology and experiences that have not been defined for the public until recently. Development in this field has been so swift, new iterations and nuances are emerging faster than we can name them. As a pioneering industry, it is the responsibility of those who are part of it to foster transparency and collaborate to create the resources we need to reach a common goal, which is adoption.

When defining concepts within immersive technology, most people defer to the virtuality continuum. Introduced by Paul Milgram and Yoko Koshino in 1992. It describes the range of realities between the physical and the virtual as Mixed Reality (MR).

By Stephanie Llamas

Stephanie Llamas is Vice President of Research and Strategy and head of immersive technology insights at SuperData Research. She is a widely respected thought leader frequently quotes in the press, and invited to speak at prestigious conferences like CES and SXSW.

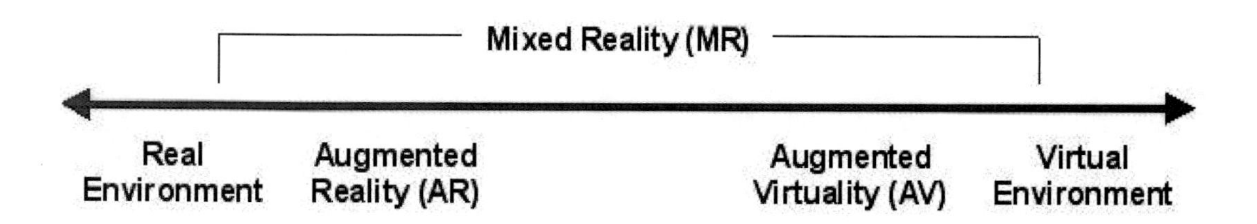

In truth, this is a relic of a different era, before we could actually conceptualize and experience modern XR (the current blanket term nearly everyone is using for immersive technologies, although its academic definition referred to bio-augmentation). These concepts do not span a continuum, but instead are built on both symbiotic and distinctive elements. 360 degree content is essential to both VR and MR since each relies on some level of immersion that incorporates a user's ability to interact using the real life space around them. AR and MR, meanwhile, both require that a user sees a real life environment. MR is a mixture of key elements required for VR and AR, but still provides a distinctive experience and, therefore, is a distinctive concept. Their unique offerings, but overlapping attributes, make them independent parts of a shared ecosystem known as XR.

In 2013, Oculus Rift's DK1 crushed its Kickstarter goal of $250K by raising over $2.4M. This drove renewed interest in virtual reality, which had traditionally been the stuff of science fiction. Once Facebook acquired Oculus for $2B in July 2014, excitement turned into a frenzied urge to definitively decide if VR was the real deal or a passing fad. The world was stunned when one of the biggest technology companies in the world paid more than 800 times what Oculus was able to crowdfund just one year earlier.

Slowly, VR, AR and MR brought in more and more stakeholders. Stakeholders with attention grabbing names like Apple and Google. With astronomical funding, hyper-secretive MR headset maker Magic Leap has raised $1.4B without a public product.

I am a strong believer that calling the game

"AR"

is about as accurate as calling two slices of bread a sandwich.

But an increase in stakeholders means an increased need for clarity. Solely using the term VR doesn't cut it the way it did five years ago. To most, it was a catch all for AR and MR, and since the latter hadn't revealed too many viable products, delineating terms weren't really necessary at the time.

Then came Pokémon GO. I am a strong believer that calling the game "AR" is about as accurate as calling two slices of bread a sandwich. Some of the pieces are there, but it is missing fundamental elements (i.e., digital images in the game do not detect and interact with the real world, which is required to augment reality). As a passionate rebuttal to this misnomer, and, even worse, attributing its success to AR, I wrote that this deceptive definition was like calling an IMAX screen a color TV, as it degrades the public's perception of how grand an IMAX screen actually is and, thus, stifles its potential growth. After all, why would anyone want to pay to go see something they have at home? Better yet, why would anyone want to use AR in other apps if the majority of Pokémon GO players turn AR mode off anyway?

And now, I am regularly embroiled in debates as to what virtual reality vs. mixed reality vs. augmented reality are, especially after Microsoft's inexplicable marketing messaging that calls its occluded, fully immersive headsets: MR. That, to me, is the definition of VR, especially since the consumer's experience in the headset is fundamentally the same as the ones they have in an HTC VIVE, Oculus Rift or Playstation VR. In the meantime, Microsoft also calls its HoloLens an MR device. A headset that actually does fall in line with the definition of MR.

I disagree with the idea that we shouldn't clearly define these terms and preach their definitions, especially since I've met a significant number of people who do work in these industries, but still have little understanding of the differences between the three. This is particularly important as we try to define a technology that will eventually be made for a mass market.

XR adoption can only happen if people value using it. Consumers do not research new technology. They are told what it is and how to use it in a simple enough way that they can evaluate its potential value to them. The concept of smartphones was easy to understand because people already knew what its key elements were (mobile phone and the internet) and its unique value (convenience and improved user interaction). VR, AR and MR offer distinctively different experiences and right now the industry suffers from a perceived catch all. For instance, if people perceive VR to only be a small step away from AR, why would they spend hundreds of dollars on a headset for a marginally better experience when they already have an iPhone that can do mostly the same thing?

So without further ado, here is a taxonomy that makes sense:

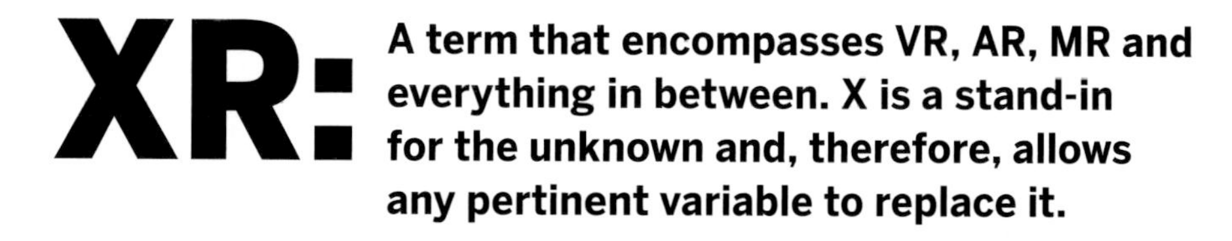

Trying to pull further meaning from the X negates its actual meaning. For instance, Qualcomm coined the term "extended reality", using the X as a semiotic sign, standing in for something specific. But the X is actually meant to do the exact opposite. It is standing in for different variables, and simultaneously nothing in particular.

While that explanation is stuffy, complex and academic, it's the only way to show the differences between a term that comes from a long understood definition (X = unknown variable). By fighting against the status quo, Qualcomm has decided XR pertains to its mobile ecosystem, further adding to a long list of difficult to decipher terms often used in the wrong way by people who ought to know better.

It is not the right or privilege of companies like Qualcomm or Microsoft to create their own vocabulary that goes against agreed upon definitions in order to differentiate themselves on a marketing level. And in such a nascent industry, it is the responsibility of every stakeholder to reach a consensus on the most important aspects that will bring in consumers, which starts with a common language.

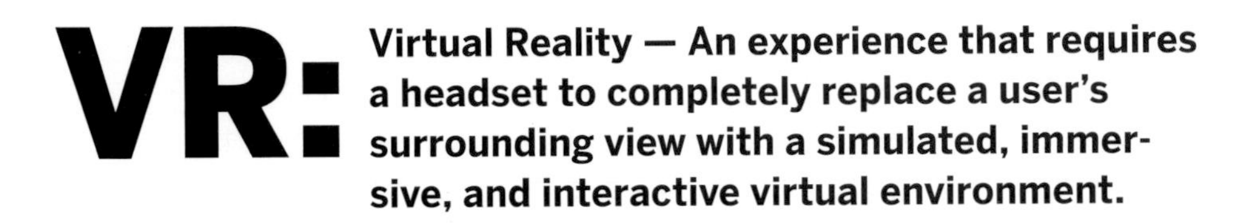

VR: Virtual Reality — An experience that requires a headset to completely replace a user's surrounding view with a simulated, immersive, and interactive virtual environment.

The consensus around VR's definition is the strongest among the three realities. Because it sparked the XR revolution, people had early access to VR. Although it remains difficult to explain to consumers who have never tried it, once they have, there is some semblance of a consensus, even if they still need some extra education. Basically, we can all agree that in order for something to be VR, it needs to completely replace the user's actual reality with a virtual environment.

The levels of agency required to create VR is a debate as to how immersive and interactive something needs to be. For instance, the simplest form of 360-degree videos do not allow users to interact with or change anything they see, only the direction of their gaze.

Excluding 360-degree video because it is not interactive is incorrect because fully immersive 360 degree video, which completely occludes a user's real environment, will always give them agency. Unlike a 2D video where your field of vision is dictated by what the director can fit within a rectangle, a 360-degree video viewer can explore their surroundings however they want, creating their own unique experience with the content. Their agency is in their ability to explore all perspectives on their own and come away with the experience they chose to have.

AR: Augmented Reality — Overlaying or mixing simulated digital imagery with the real world as seen through a camera and on a screen. Graphics can interact with real surroundings (often controlled by users).

Augmented reality is distinctive from VR and MR in that it cannot be immersive. Immersion can only really be achieved if the technology follows a user's vision, something that currently can only be done with glasses or a headset (and even glasses can restrict field of vision, affecting immersion). AR currently relies on a screen that users move with their hands, not their head (e.g., a smartphone). Therefore, they are once removed from their augmented reality, not existing within it, but seeing it through a digital window. Actual reality is required for AR experiences, but the screen's frame makes it impossible for a person to be present within a 360 degree environment.

Mixed Reality — An experience that always gives the user a view of their real surroundings, but uses a headset to overlay graphics that are interactive with actual reality (augmented reality), and/or incorporates elements from actual reality into a virtual environment (augmented virtuality).

MR fills in the differences between AR and VR, surrounding a user with virtual imagery and, therefore, providing immersion. It has the ability to incorporate and allow interaction between all realities so that they can work symbiotically to create the ultimate convergence of the real, virtual and augmented.

This is why Microsoft's appropriation of the term MR to describe an occluded headset that completely immerses the user in a virtual reality is a marketing message that is in fact hurting the industry. The general public will never understand the nuance within the definitions above, nor should they have to. The only way to really clarify the definitions is by showing the differences in experiences and, thus, taxonomy.

When you get into a VR headset, you understand what VR is. Looking at AR on a mobile device shows you what defines AR. And getting into an MR device that melds different elements of XR to create an alternate reality gives the user a first hand understanding of what a mixed reality feels like.

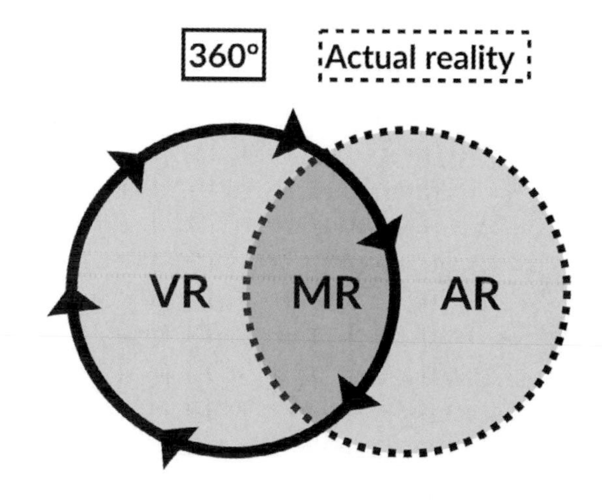

So when a user puts on a Windows MR headset and has much the same experience they would in any other VR device, it's difficult to make a distinction. And if Microsoft dubs Windows MR and the HoloLens both as MR devices, does that mean they are the same? This has created confusion for the industry and consumers, and it even seems Microsoft is confused.

On their HoloLens website they say "This is mixed reality. Interact with virtual holograms and real objects in your physical world with holographic technology enabled by Microsoft HoloLens."

But then on their Windows MR website they say, "Windows Mixed Reality combines the thrill of virtual reality mixed with augmented reality in gaming, travel and streaming." So which is it? Virtual holograms overlaid onto the real world, or a combination of virtual and augmented realities?

Yes, this taxonomy is confusing, nuanced and dense. However, the words themselves are not meant for the consumer. It is the industry's responsibility to understand these terms and show their differences to consumers in a way that does not confuse or obscure the technologies' fundamental meanings. Without clarity, it is impossible for consumers to weigh the value propositions of each technology, or understand the relationships between them in order to make informed purchasing decisions.

It can also degrade reputations. For instance, if someone thinks mobile AR is just as good or fundamentally the same as VR or MR, there is no need for them to expand outside their smartphone's screen. Dissention hurts the market's potential. Something that all XR companies should be united in trying to grow. Coming together as an industry can only be done through transparency and consensus. Without them, the only thing that will grow is the gap of disappointment. ■

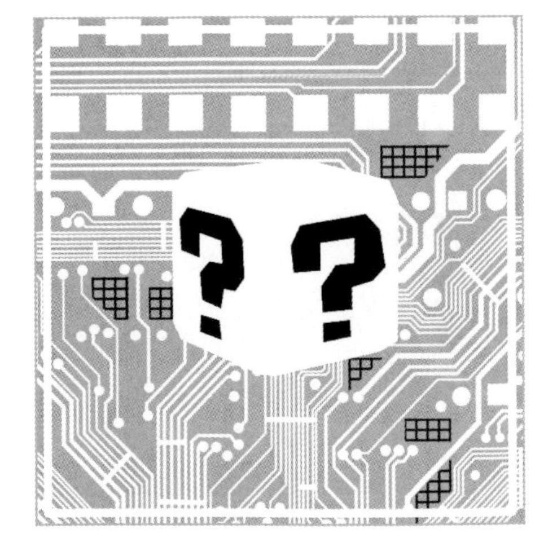

The general public will never **understand the nuance** within the definitions above, **nor should they have to.**

AR/VR War Of

By Charlie Fink

Charlie Fink writes about VR, AR and New Media for Forbes. In the 90s, he was EVP & COO of VR pioneer Virtual World.

Virtual reality is about humanity's quest for immersion. It provides presence and agency in other worlds, in stories and myths, and it stretches from Plato's cave to religious rituals, theater, dark rides, theme parks, film, television, and video games. These experiences require our willing suspension of disbelief.

Augmented reality, on the other hand, has its historical antecedents in tools. Humanity has always sought tools to make people stronger, faster, and smarter. AR is the ultimate expression of man's quest for mastery. It is a tool, like a club. This is why it's seeing its first real applications in commercial circumstances: Beat your business with this club and money pours out.

In short, VR is a new reality. AR is about enhancing reality.

Current dogma is that VR is an extension of AR, as represented by the Milgram Scale (aka the Windows' Mixed Reality (MR) Continuum). It is no such thing. VR and AR come from different places and they seek to do different things. New and enhanced are not the same, and they are not on a continuum. They use some of the same technology, and 3D objects and AI are important to both, but so what? VR and AR devices are computers, and computers do a lot of different things.

In 1994, Professor Paul Milgram and Fumio Kishino created the mixed reality spectrum to explain the relationship of AR and VR. On one end is physical reality, and on the other end is a fully occluded, digital world. They described their definition as "a line between the extrema of the virtuality continuum, which extends from the completely real, transits augmentation, and arrives in the completely virtual."

Words

MR/XR

The $3,000 Developer Edition of Microsoft's HoloLens is the first commercially available augmented reality headset to see wide enterprise adoption.

Courtesy AWE

Microsoft Mixed Reality Spectrum

Mixed reality blends with physical and digital worlds to produce new environments where physical and digital objects co-exist and can be interacted with as if all of them were real.

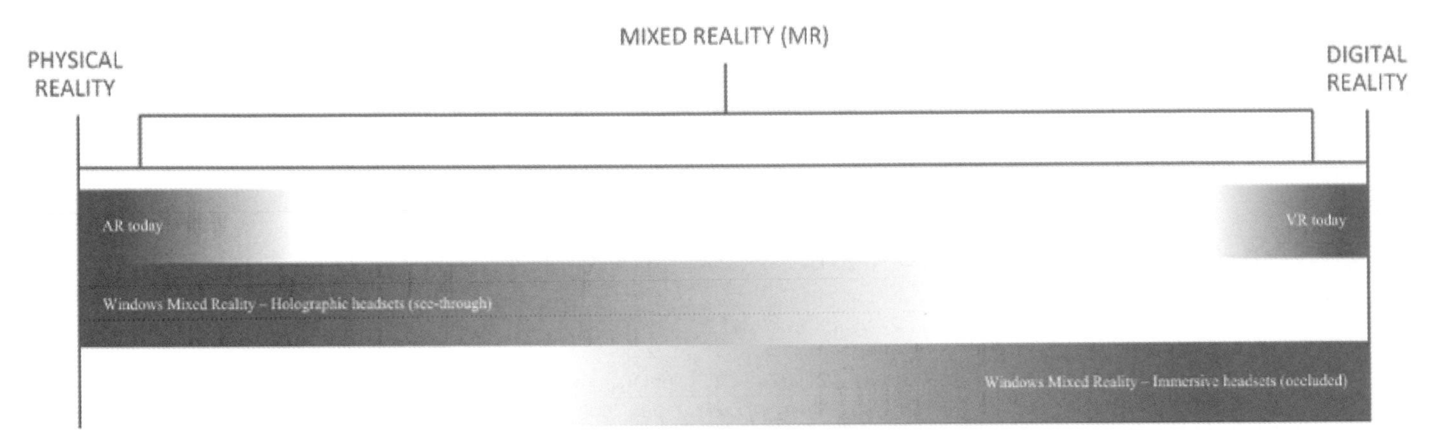

Earlier this year, Microsoft introduced Windows 10 MR and trotted out an updated Milgram Scale, rebranded as "The Mixed Reality Spectrum", in which the HoloLens and their new occluded VR headsets sit conveniently on opposing sides. This is wrong, and not just because they don't credit Milgram.

AR head mounted displays (HMDs) are inevitable, and undoubtedly there will be headsets on the market that do both VR and AR, but that does not mean these activities are the same.

The quest for immersion represented by VR, and the desire to be augmented, vis-a-vie AR, are very different things. Until now, academics and scientists have seen VR and AR related on the spectrum, with one side anchored in physical reality and the other in the fully digital virtual world. This has been in most of my presentations, until I had an epiphany. It's not true. One doesn't lead to the other. The hardware distinctions may at times be subtle, but the difference in use cases is not.

Similarly misleading is the notion that the more immersive the experience, the more VR-like AR becomes. I cannot think of a single use case or proof of this. You are either augmenting or replacing.

I like the people at Microsoft. Their appropriation of "MR", as in "Windows MR", should be kind of funny, as they are stealing the flawed Milgram scale. It's the perfect bank job, until the loot turns out to be monopoly money.

Just in case we weren't confused enough about the language, Microsoft is introducing Windows 10 MR (MIXED REALITY) with new fully occluded VIRTUAL REALITY headsets. Just let that sink in for a second. Raise your hand if you are NOT confused.

To make matters worse, the Consumer Technology Association (CTA), of which Microsoft is a member, put out a set of AR/MR/VR/XR definitions last fall, which it and apparently the rest of the industry, appears to have ignored. I would like to say they tried, but if a press release falls in the forest and no one hears it, did it make a sound?

Also, as long as we're at it, Microsoft has also unhelpfully confused the definition of holograms by dubbing the HoloLens, a "holographic computer". Seriously?

Note that the key quality of the hologram is that it can be viewed by the naked eye *without the use of a headset.*

I would be remiss if I did not call out Qualcomm's appropriation of the term XR, defined for decades in this way: "X Reality (XR) consists of technology-mediated experiences that combine digital and biological realities". Qualcomm's website, and a new secondary Wikipedia post publicly questioned by Wikipedia's editors, defines "Extended Reality (XR) is an umbrella term encapsulating Augmented Reality (AR), Virtual Reality (VR), mixed reality (MR), and everything in between".

The good news is that the war is not over. Even if Qualcomm were to succeed in redefining XR, and even if granted the trademark, they could never defend it in court. With that in mind, they are using a common workaround, identifying XR with a graphic mark registered with the US Patent and Trademark Office. #Resist ■

Right now, the largest companies in the world are preying on this confusion like a pack of hyenas.

The AUGMENTED Man

If you were halfway to work and you realized you left your cellphone at home, would you turn around to get it, even if it meant you'd be late?

By Charlie Fink

It is strange to see the old movies and TV shows where characters use landline phones, especially pay phones. It reminds us that so much more than fashion has changed. Time and space are different. People gave directions and used paper maps. We looked everything up in paper encyclopedias, which were expensive and kept in public buildings called libraries. We made idle talk with strangers at bus stops. At the same time, something deeper changed in the way we relate to each other, disrupting even the relationship between parents and children. We are less present, but available on demand.

When I started out in the movie business in the 1980s we used fifty year old technology. The telephone was the only tool most businesses had. In fact, millions of people were employed "to handle the phones." Americans went to the moon on the slide rule. Yet, less than 100 years before, the vast majority of humans never traveled more than fifty miles from where they were born. Men still owned other men.

In twenty generations, we have gone from pre-industrial serfs to star traveling magicians. In the context of the 125,000 generations that came before us, that is astounding. It is the ultimate hockey stick.

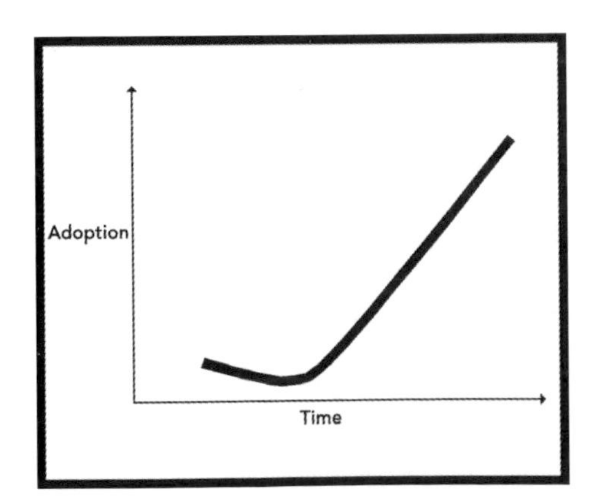

Now we each carry a $1,000 brick "brain" that contains our most valued possessions. Though the content may be stored in the cloud, the access to it, the interface, is in our hands. We look into our palm hundreds of times a day. It's hard not to. It has access to all the information in the world and to everyone we know. When was the last time you didn't know something?

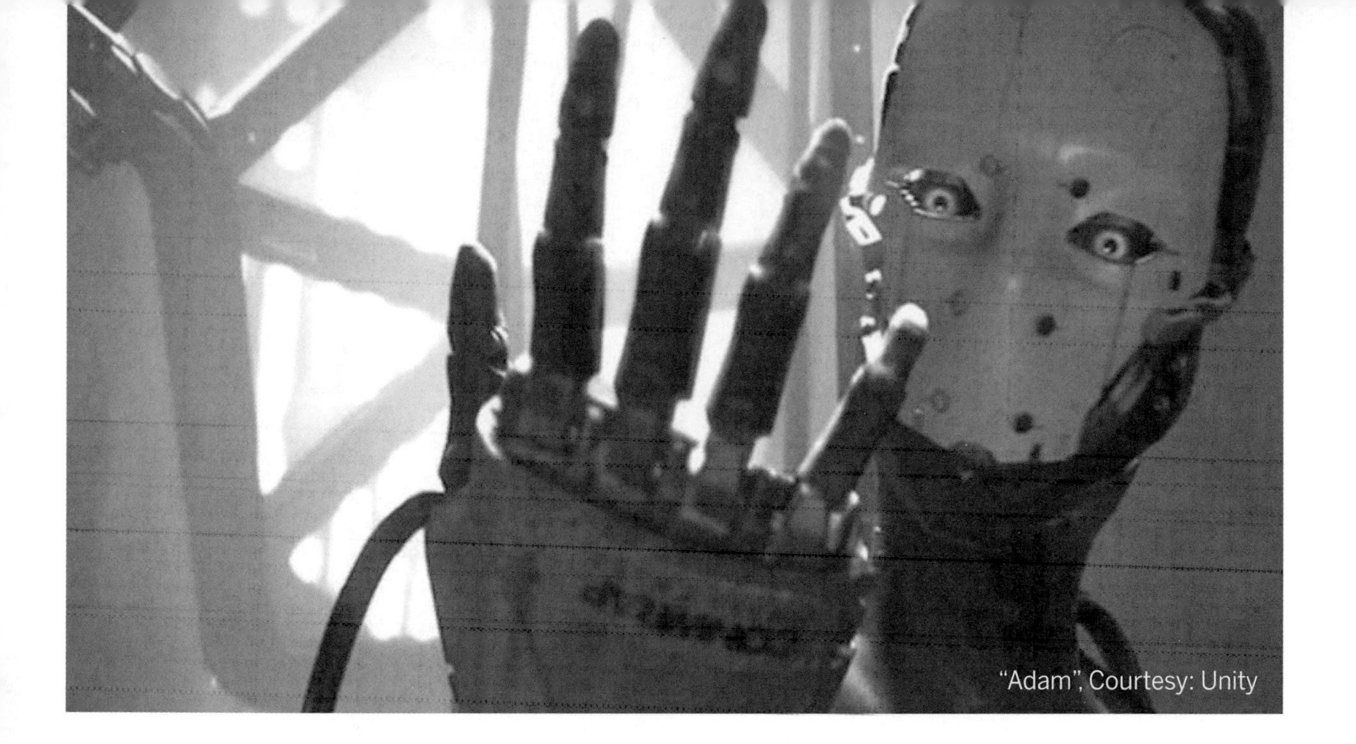
"Adam", Courtesy: Unity

We are constantly shifting between the world in our hands and the physical world around us. Thousands of people die every year because of this incredibly bad form factor, and the need to change it is urgent. It took 15 years to dig this hole, and it will take 15 years to dig out. But when we do, it's going to be a very different world, because advances are happening at many other levels. AI, the blockchain, cloud computing, are also rapidly changing computing. And then there are the unintended consequences of our greatest inventions, like Facebook, which may better connect us, but also make us less free.

We are multitasking while continuously interrupting ourselves by checking our smartphones a hundred times a day.

Pattie Maes of the MIT Media Lab influenced my thinking about AR and how this tool is going to change humankind. Well known for her compelling TED talks, she gave a mesmerizing big picture presentation on the state of modern life: Living simultaneously, imperfectly and superficially in the real and digital worlds. We are multitasking while continuously interrupting ourselves by checking our smartphones a hundred times a day.

"Tech has become part of us," she said, "and we all know there is no going back. Instead, we need to make our integration with technology more seamless. In the short term, our technology is making us less mindful and attentive. But in the long term, it will make us much, much better.

Better learners, better workers, better able to reach our potential. We will become augmented. Cyborgs." Many of the Media Lab's forward thinking projects seek to accomplish human augmentation with wearables, biometric scent dispensers, and even tattoos. I always thought VR meditation was utter new age bullshit, but when Maes showed how we can use brainwaves to levitate and move objects in a 3D digital world, I shed my cynical disbelief.

Most mobile phones already do a form of augmented reality, expressed in the simplest form, with stickers and filters that can be mingled with the real world. Apple's new operating system enables the phone's camera to become the interface, a window through which we view a world where the real and digital mix. But holding your arm out is ridiculous, unnatural and uncomfortable. It's the worst form factor ever accidentally invented by man, and if the app is useful, it makes your arm hurt. It's not sustainable.

"A head mounted display would be just a waypoint, as we will continue to evolve beyond it on our march toward human-machine integration," Maes told me in a recent email.

The goal is to get this information into the brain. Right now, the eyes and the ears are the only ways into the brain. "Computing interfaces will become more and more invisible to the user/wearer" says Will Schumaker, a postdoctoral scholar at Stanford University, whose focus is optics. "But going beyond HMDs will be tricky and require quantum leaps in optics or biotech, which might take some time."

Schumaker says bio-augmentation is more than science fiction, though still far from science fact. "Digital contact lenses or interocular devices that might be directly implanted with a lasik-like procedure require technology that hasn't been invented yet. There's been some work with contacts, but everything has to go into a wafer the thickness of an eyelash. I think they can get a Twitter feed's worth of data into them."

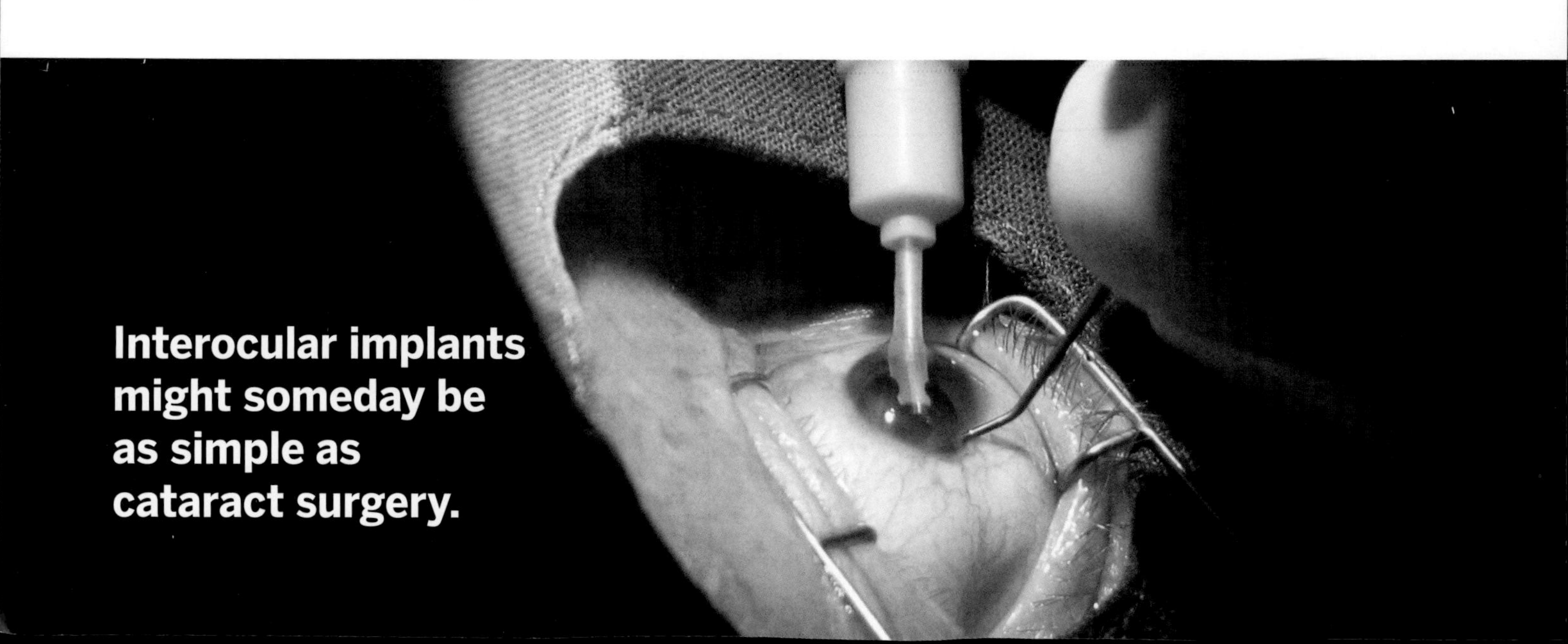

Interocular implants might someday be as simple as cataract surgery.

Think about how useful this could be when, not if, the form factor could be solved, and if the data in the lens could geolocate and then seamlessly negotiate contextually relevant information with social networks, literally right before our eyes. A digital AI, perhaps Alexa, will have a dimensional presence nearby on command. Everything controlled by voice and gesture. A depth detecting camera with facial recognition could allow you to identify social contacts in public places. Imagine walking down a city street. Every billboard a portal. Every storefront the store. Every restaurant rated by social contacts. Of course, it will also enable drones to track and kill everyone who likes Billy Joel on Facebook.

Neuralink Corp. is a startup cofounded in early 2017 by billionaire Elon Musk, who runs Tesla and SpaceX. He recently disclosed he has invested $27M in Neuralink, a company that implants electrodes into the brain in order to create a "neural lace", the ultimate man-machine interface. It may be decades away, but it's not as crazy as it sounds. Medical devices are implanted today to treat certain diseases like Parkinson's.

Of this we can be certain: Our marriage to the handheld smartphone, and the terrible form factors associated with it, will eventually come to an end. Headsets, as fashionable as Snap Spectacles, will be capable of much more. Handsets will do less. All that may be left is a Bluetooth keyboard. The 3-year mobile upgrade cycle will continue. Augmented reality glasses, like the smartphone, will come to you in your upgrade after the next one, and the next one, and maybe the next one after that.

Man will be augmented. Man will merge with machines to become cyborgs. Always on. Always connected. Continuously updated. It sounds like science fiction, but it's already happening. It cannot be stopped. It cannot be timed. It cannot be resisted. It might not be all good. "The law of unintended consequences can be profound in this area," Schumaker warned, "the benefits of this technology will not be shared equally." ■

THE QUEST FOR

By Charlie Fink

While AR is a tool that augments us, makes us faster, better, and smarter, VR does something very different. Think of it this way: AR is augmentive; VR is transportive.

VR changes the world we're in and we accept the illusion eagerly. While watching television, we desire escape into imaginary worlds so intensely that we forget about our surroundings and the laws of physics. Think of the millions watching movies on their phones at this very moment. Despite the tiny screen, they are able to block out every shred of reality and suspend their disbelief, focusing on a miniscule fraction of their field of view to the exclusion of all else. The brain doesn't have to work so hard in VR, though it is as dependent on our suspension of disbelief as its predecessors: Theater, fun houses, movies, and television.

In VR, we seek presence in virtual worlds once promised by religion. Unlike AR, we do not need or want a tether to physical reality. One might argue VR is, then, metaphysical. It is reaching inside us for something that is not a tool, but our soul.

We see the origins of these desires for immersion in humankind's earliest days. Stories, told around campfires and later illustrated with pictograms, stimulated the imagination and activated a quality unique to human beings: The ability to suspend our disbelief allows us to believe impossible, magical things and see fantastic dimensions invisible to the naked eye, like Mt. Olympus, or heaven and hell.

The earliest origins of theater are to be found in Athens where ancient hymns were sung in honor of the god Dionysus. These hymns were expanded into choral processions where participants would dress up in costumes and masks. Soon, people portraying the gods performed their stories on stage. Audiences were mesmerized by a new and abstract emotion: Empathy, the connection to a fictional character. It was doubtless as new and impactful as VR is.

Around the same time, dramatic illusions like the organized shadow play in Plato's Cave illustrated myths and stories of the gods and their

IMMERSION

supernatural powers in a more abstract way, one not bound by the limitations of the physical world. At first, the ancient Greeks thought these simple but powerful tools could cure the criminal and/or mentally ill mind. Of course people knew these were simply illusions, but they allowed themselves to believe them in order to experience the story, allowing themselves to be entertained without resistance or self-consciousness. The stagecraft disappears so we enter the world of the story. This is the essence of the willful suspension of disbelief.

Humankind has always been clever about creating other worlds. The concepts of heaven and hell have captured the imagination for centuries. Visions of the after life were one of the keys to controlling ancient empires. These secrets can be glimpsed in medieval churches and modern dark rides, like Disneyland's *Pirates of the Caribbean*, which transports us to another world. We drift past, like a barge full of ghosts, mesmerized by characters locked in their stories.

The recent explosion of immersive live entertainment can be credited to the trendy "escape room" concept. Entrepreneurs believe this is proof positive there is a demand for VRcades and other out-of-home immersive entertainment. Wildly popular among millennials, escape rooms charge visitors as much as $50 per hour to cooperatively solve puzzles in order to "escape" from the room they paid to be in. While many have been taken by surprise by the escape room phenomenon, creators of games, live theater, and virtual reality experiences, in particular, have found validation, and a new audience, through the escape room.

Paradiso is an escape room in New York City that combines high tech devices with live role playing actors, some of whom pretend to be FBI agents infiltrating the cultish Virgil Corporation, including an unfortunate agent in a tuxedo who appears to be dying. For a little extra money, *Paradiso* and Virgil will follow you home to give you an intense, personal, real life experience, as the fictional (or is it?) cult tries to sink its teeth into every aspect of your actual life. *Paradiso* Director Michael Counts thinks immersive entertainment has reached a tipping point. "We are seeing the beginning of an experience economy," he told me.

Counts said live experience entrepreneurs are simply building on the pioneering work of Walt Disney. "He was the first person to realize that people want to be transported into the world of a movie. He used the technology he had at hand to achieve his vision." Put another way: You don't need to be wearing a head mounted display to be fully immersed and present in a fictional world.

Then there's *Sleep No More*, a massive immersive theatrical experience where guests mingle in a wordless, multi-floor Edwardian treatment of

"I've seen Disney animation storyboard artists bring a single drawing to life with a pointer. They created Academy Award-winning performances tailored entirely to an intimate production team rather like cavemen spinning a good yarn to accompany the drawings viewed in the cave's firelight.

The epic success of "Sleep No More" in New York represents a tipping point for immersive entertainment.

photo credit: The McKittrick Hotel

MacBeth inside the old McKittrick Hotel (it's actually a converted warehouse) in Chelsea, a neighborhood of New York City known more for speakeasies than theaters. A UK import, the show has been a hit since it opened to rave reviews in 2011.

Sleep No More is now going into its sixth year. The actors perform the story in pantomime across a very large, multi-floor building and the audience members must decide for themselves who or what to follow, and the best perspective for doing so. Unlike *Paradiso*, which casts you as a character in a developing drama, in a deconstructed narrative like *Sleep No More*, it is extremely difficult to suspend one's disbelief and fully enter the MacBeth narrative or feel empathy for the characters. For this reason, the producers seek to immerse customers in the larger McKittrick Hotel mythology and have them wear Venetian costume masks, like the characters in Stanley Kubrick's demented masterpiece *Eyes Wide Shut*.

The at-home VR experience of *Sleep No More* might be so bandwidth intensive it would take many hours, even days, to download with a cable modem. Penrose Studios' new volumetric animated VR experiences, also pantomimed (which animation does so well), *Allumette* and *Arden's Wake*, offer clues as to how a "ghost" audience within the virtual world might have a better perspective and emotional connection than to a traditional 2D story, which has the advantage of cinematic language to communicate emotion.

In Penrose's VR experiences, we have the ability to change our physical perspective on the characters and narrative, but we cannot change or affect the action or the world we occupy with them. We can put the characters in the palm of our hands, follow them into mouse holes, and generally be with them in the world. Not being able to help the little match girl right in front of us makes the end of *Allumette* all the more heartbreaking and does so in a way traditional narrative storytelling never could.

Today most VR experiences are limited. Even the most realistic digital renderings are visibly just drawings. There's no AI. Content is stubbornly linked to games, although there are many practical applications being adopted by enterprises and the military, particularly for training. It's fairly common to see cartoonish worlds, illustrated worlds, or 360-degree video (which is not really VR) in a VR headset. Far less common are photo realistic worlds captured through photogrammetry. Within them, our avatars can explore the environment and interact with other human driven avatars.

In the HBO series *Westworld*, AI characters compete with humans to draw us into a story which appears so real, characters question what is real and what is a simulation, and can physically die in the story. While far from reality today, in the the next ten years expect such a VR experience to attract millions. This killer app of social gaming will almost certainly be based on a significant intellectual property, like *Star Wars*. If true, the four BILLION dollars Disney paid George Lucas will be among the biggest media steals of all time. If I am wrong then some other franchise, perhaps *Westworld*, or the Spielberg movie *Ready Player One*, whose principal action takes place in an infinite virtual world called "The OASIS[1]", may be the story to follow.

Presence and agency in a photorealistic simulated world is the holy grail of immersion. In the meantime, suspension of disbelief allows us experiences that obscure the line between what's real and what's virtual. ■

[1] OASIS (Ontologically Anthropocentric Sensory Immersive Simulation).

PRESENCE AND AGENCY

in a photorealistic simulated world is the **HOLY GRAIL** of immersion

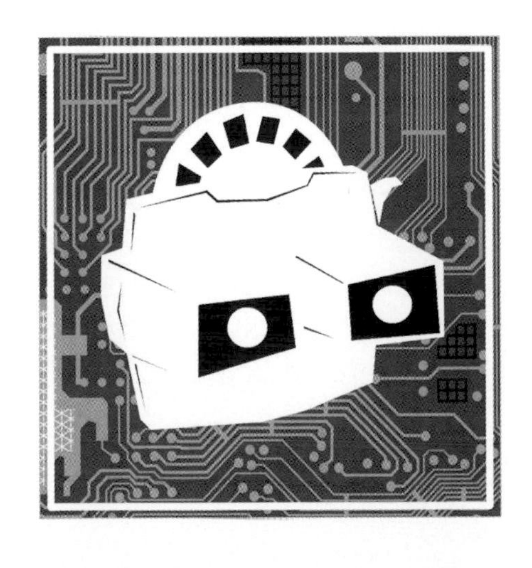

POP CULTURE Predicts The Future

By Michael Eichenseer

Michael Eichenseer is a writer for VR Voice, exploring VR and its affects on humanity. He is most interested in game design and human behavior.

Artists deal in both the possible and impossible. Many, especially science fiction writers, play somewhere in the middle. They write about technologies that don't yet exist. Or they assume the possibilities of technologies far ahead of their time.

Da Vinci's sketches, writings, and musings were at times centuries ahead of possibility. It's not an overstatement to say we can attribute the entire aerospace industry to his vision. Da Vinci began a tradition of futurism in art and science, and made a strong statement about the relevance of the relationship of imagination, art, and science. It is the ability to conceive and share a vision, to make the invisible visible, and therefore possible.

This art of futurism has been practiced by novelists, who two hundred years ago created the science fiction genre, beginning with Mary Shelley's *Frankenstein* in 1818. It is no accident that science fiction is every technology geek's favorite genre. When a geek's imagination gets combined with their knowledge of technology, science fiction can, and has, become science fact. Thanks to megahits like Star Wars, Star Trek, Alien, Avatar, The Terminator, The Matrix, and Guardians of the Galaxy, science fiction is well entrenched in popular culture. Through each representation of virtual reality and augmented reality in the past, writers and artists inspired technologists and entrepreneurs to create new things.

Some companies today attribute their success and core technologies to the inspiration of science fiction. Dreamscape Immersive is using free roam VR

credit: Joe Prohibition

to create an experience very much like the Star Trek holodeck, created to relieve the relentless monotony of space travel. VR Games like Lone Echo pit two teams against one another in a zero gravity arena, influenced in no small part by Orson Scott Card's 1985 novel *Ender's Game*.

The influences of popular culture on AR and VR run deep and will continue to do so. Philip Rosedale, Founder of the massive virtual world Second Life and the recently opened High Fidelity VR metaverse, said he was so inspired by the movie *The Matrix*, he stepped down from his role as CTO of Real Networks to start Second Life. "I said 'I'm going to make that, but it's not going to look like that," Rosedale recalled in a recent interview.

Snowcrash: Welcome to The Metaverse

The term Metaverse was coined well before it was popularized by Neal Stephenson in his seminal 1992 novel *Snowcrash*. The word "metaverse" combines the prefix "meta" (meaning "beyond") with "universe", here referring to an infinite number of interconnected virtual or digital spaces. Stephens on's book was more than just prescient, which is why he's been called "the tech Nostradomus". *Snowcrash* has influenced thought leaders and inventors for twenty-five years, including Jeff Bezos, Google Earth designer Avi Bar-Zeev and Magic Leap founder Romy Abramovitz, who hired Stevenson as a Futurist.

Ernest Cline's novel, *Ready Player One*, also describes a dystopian future where many people prefer a virtual world, called the OASIS, to the real one. "For my money, *Snow Crash* is still the greatest virtual reality novel ever written, and it had a profound influence on me when I was writing *Ready Player One*". Cline wrote in an exchange with Charlie Fink on Twitter. "The Metaverse is a perfectly realized visual world that Stephenson crafted with a programmer's eye for detail, and it directly inspired the OASIS in my own novel." In 1992, Stephenson predicted a world remarkably similar to the one we're living in today (remember, everyone underestimates the future) featuring smartphones, GPS, VR and AR HMDs, crypto currency and the gig economy. Stevenson is also credited with bringing the Hindu term "avatar" into popular culture.

"There's so much to love about this book," Cline continued. "The tech-savvy rapid-fire prose. The super cool computer hacker / master swordsman / pizza deliverator main character named Hiro Protagonist, and the crumbling future America that serves as its setting. It's still as much fun to read as when it was first published a quarter of a century ago." Hiro, who makes a crazy living working an Uber type gig gets warned off Snow Crash (a street drug named for computer failure) by a batty acquaintance, who gives him a file labeled "Babel". Complications ensue.

The Matrix: Do you think that's air you're breathing?

In 1999 *The Matrix* downloaded to theaters everywhere. *The Matrix* depicts a future where machines have taken over the surface of Earth while most humans are "wetwired" into a system known as The Matrix: A computer simulation of the real world so real its inhabitants are unaware of the outside world.

Users in the matrix have surgically implanted "ports" near the brain stem, down the spine, and on each limb. Tubes implanted throughout the body provide the nutrients necessary to keep the human alive. The unplugged character Morpheus refers to this as being wetwired.

With direct wired connections to a person's nervous system, The Matrix simulates the same responses on a user's nervous system as experienced in reality. Those jacked into The Matrix don't need to wear special suits or wear heavy head gear. Users lay in a chair where a cable can be plugged into a socket embedded in the back of their head. Software depicts what the user sees, feels, and experiences.

VR enthusiasts hope for a level of immersion as real as The Matrix, but fear the dystopian future of machines trapping the entire human race within a simulation. Direct neural interfaces are a desired future of many VR enthusiasts, and the required technology is a hard problem to solve given the complexity of the human mind. Futurists, such as Ray Kurzweil, believe human brain computer interfaces are inevitable within the next two to three decades, while others doubt they will ever be feasible.

There are less invasive forms of neural interfaces in pop culture. A recent phenomena, *Sword Art Online*, a Japanese anime series, presents a technology called Nerve Gear. Nerve Gear is a head mounted display users wear, usually while lying down. The head gear interfaces with a user's nervous system and simulates the feelings of the virtual world directly to their brain. This way no heavy haptics hardware is needed.

With the nerve gear on, users can immerse themselves in a digital world from head to toe with all their senses. Their real body is in a comatose state in the real world. They can only be in VR for a limited time before needing to go back to reality to attend to their bodies and real lives.

Avatar

In James Cameron's Avatar, a crippled marine is sent to the planet Pandora where humans are mining a precious metal known as Unobtainium, which is apparently valuable, but extremely difficult to obtain. To do so, they must destroy the habitat of the native species, the Na'Vi. As a means to communicate and understand these natives, scientists have created "Avatars". These avatars are biological copies of a Na'Vi that human pilots can jack into and use to meet and greet the aliens on their own terms. The presumed Bluetooth type connection between the puppeteer and the avatar is never explained.

Jake Sully lays in a pod and has his mind uploaded into the Avatar's body, in which he can walk again. The avatar body has working legs. His physical body does not. When we are in VR, we are piloting a virtual body in virtual space. The only difference between the marine piloting the Na'vi, and you being a ghostbuster at The Void, is the headset.

Star Trek's Holodeck

Star Trek's Holodeck first appeared in 1974 and depicted a near perfect simulation of reality. A room used for entertainment, education, training, and all sorts of plot twists. It has been a VR enthusiast's dream for decades now to create --- a perfect holodeck like that in the Star Trek series. Tim Ruse, Founder and CEO of free roam location based VR (LBVR) says The Holodeck was his original inspiration for starting a free roam VR company.

The Holodeck is said to use microscopic projectors that projects light directly into the eyes of the room's occupants. The projectors follow the individual eye movements and position of each occupant. No matter how many occupants there are and no matter where those occupants are standing, the Holodeck appears real to them. While eye tracking and projection technology can accomplish these features in theory, the precision required to simulate reality on the level of a holodeck appears far out of reach.

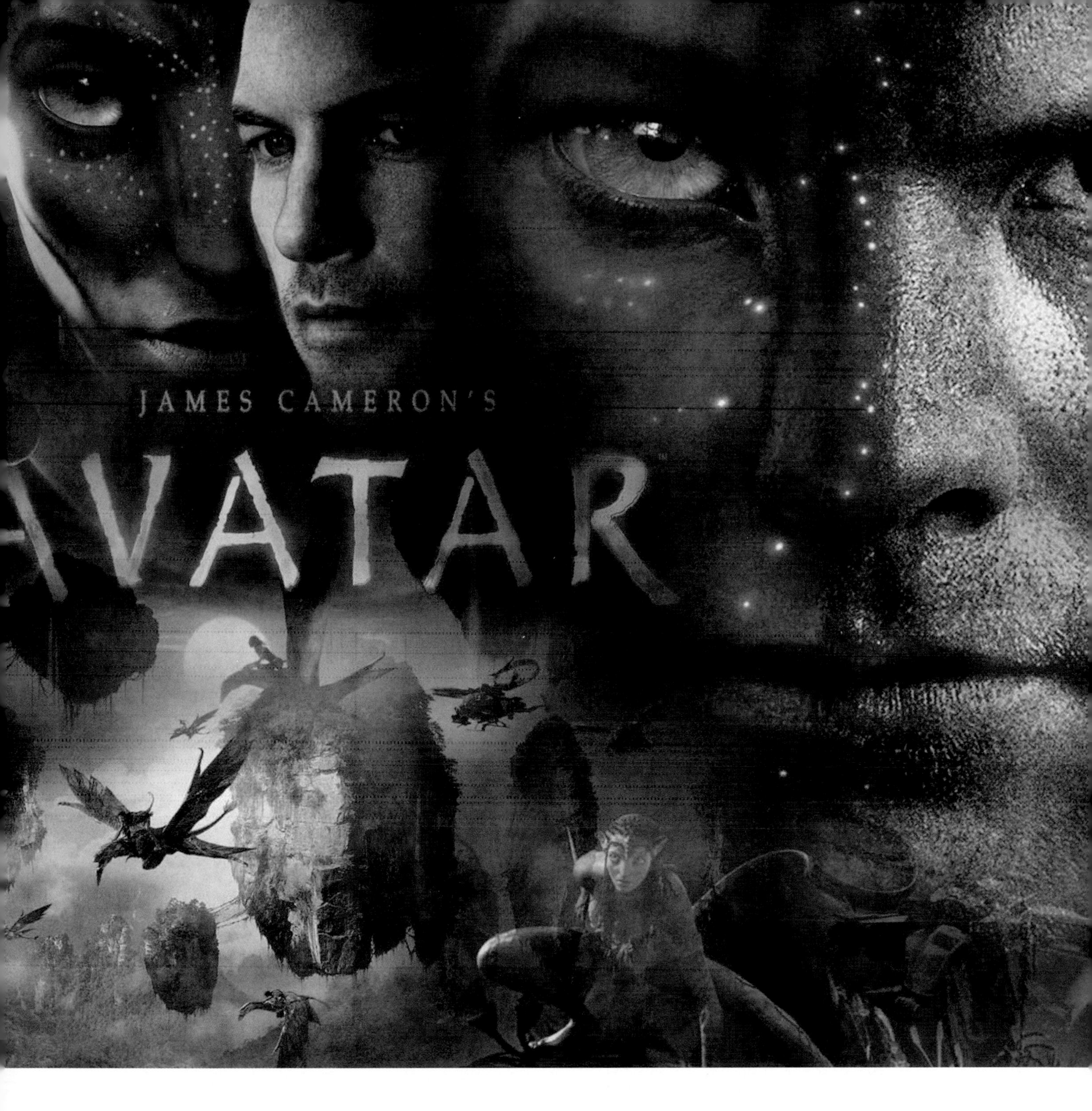

Everything about the Holodeck can be done today. The Void, Zero Latency, Dreamscape, and others (see our chapter on LBVR) are creating a universe in 30'x30' rooms you can walk around, as if real.

Star Wars

One of the delights of *Star Wars* is its inventive, unexpected revelations of high

(continued on page 47)

he Reality of VIRTUAL REALITY In Ready Player One

Ready Player One (RPO) (2011) is a prescient novel by Ernest Cline about a dystopian near future where young people, "the missing millions", spend most of their time in a metaverse called The OASIS, which also happens to be the most valuable company in the world. Imagine if Google owned not only search but every site you visited. When the Oasis' Steve Jobs-like creator, James Halliday, dies, he offers his metaverse to anyone who can solve a puzzle related to 1980s trivia. The upcoming highly anticipated movie based on the book is being produced and directed by none other than Steven Spielberg.

Almost immediately after the announcement, an industry begins to emerge around the search for clues to help solve Halliday's puzzle and produce the three keys that lead to his fortune. Those searching are called "Gunters" (a portmanteau of "egg hunter"). A massive ruthless corporation, IOI, bent on taking control of the OASIS and restricting access via a paywall, employs an army of its own Gunters to search for the keys as well.

When Wade Watts, the protagonist of RPO, discovers one of the keys, he vaults to the top of the leaderboard, becoming instantly famous. He's stalked by IOI, which attempts to recruit him, and then tries to kill him.

Wade Watts attends virtual class on a combat free virtual planet full of virtual parks and schools. He logs in to attend a class taught by a virtual teacher, presumably the avatar of a real teacher, and his access to non-educational materials is limited while logged into the classroom.

This one is an interesting choice by the author, Ernest Cline, that reflects his predictions for the expansion of e-learning. I agree we're going to continue to see enormous growth here. However, if you live in a crowded urban area as Wade does, in 25 years, it's likely in-person high schools will still persist. Advanced education and large parts of college will likely be conducted through e-learning, but it's doubtful teenagers could self-motivate enough to enjoy the benefits of online classes.

The coming purge of private universities will accelerate this trend. It may well be that college will consist of two virtual years, and two on-campus as almost no one will be able to afford $400,000 at a private university. For those around the world, e-learning MOOCs (massive open online courses) are available, often for free. This democratizes education.

Players in the OASIS can purchase their own virtual goods, like mansions, spaceships, asteroids, weapons, shortcuts and magic spells.

You can already do this in Second Life, Sansar and on the world building platform High Fidelity. Second Life has its own currency with real world value. Users buy three dimensional objects posted by creators in the Sansar and High Fidelity stores.

The OASIS also allows for the purchase of virtual goods that are interconnected with real goods, like a virtual pizza order that connects to a pizza delivery service that will deliver a slice in real life.

Amazon Fresh already does this too. Seamless and Uber Eats give you access to food delivery 24/7.

Wade Watts travels through a metaverse of connected worlds that are partly restricted by real life physics and partly restricted by a business model that relies on virtual credits. For example, Avatars have a top speed at which they can run in the virtual world. However, players can buy virtual teleports using credits that cost real money to move between worlds. They can also use technology, like virtual cars and spaceships, to move on and between different worlds in the metaverse.

The metaverse exists and is a real thing, depending on how you define it. VR enabled web browsers now make this much easier. In addition, social VR, as practiced by Sansar, AltSpaceVR, Oculus Rooms, and apps like Rec Room, also creates a metaverse of sorts, as players navigate from room to room. The creators of these worlds could introduce weightlessness if they wanted.

Virtual credits exist today. Your credit card is already on file. You'll authorize the charge, or have a virtual wallet for microtransactions, just as you purchase apps today. And then there's blockchain, which Philip Rosedale (founder of Second Life) told me in an interview last year that he is planning to integrate with his new social metaverse, High Fidelity.

Players in the OASIS have a basic headset and haptic gloves, but some higher-end systems have entire immersion rigs with treadmills to mimic movement as well as full body sensations.

These types of full body immersion rigs are available today. However, they are not integrated into many apps. Hand tracking is here. Gloves, which add haptical feedback to virtual objects including keyboards (which I think will be important) are on their way and will allow us to eliminate both the screen and the CPU.

Users enter the OASIS by verbally stating a passphrase, recognized by the OASIS system, and are allowed to remain anonymous to other OASIS users. Avatars do not need to reflect the gender, race, or even humanoid features of the user.

If you think Internet trolls are bad today, what until you're in the same room with them. Doubtless new forms of spam, ads and phishing will continue to follow us wherever we may go, in the real and virtual worlds. Anonymity and identity are big issues that VR has yet to grapple with. Rosedale thinks the solution is blockchain and is building it into his new platform. He is thinking about how your money, and digital objects, travel with you into digital worlds in a seamless and secure manner.

Players can create private chat rooms, virtual spaces only accessible by invitation.

Facebook Spaces and Oculus Rooms are only open to your Facebook friends. Oculus's landing page is like your Facebook page. Note friends list. You can configure your home screen, or room, as you like, and control who can enter. On Sansar, you could even charge them to come in.

Avatars can die. Avatars have virtual health quantified in "hit points." If you run out of health because of a fight with another avatar, or doing something that damages one's avatar, it's game over. Any items in the possession of that avatar are lost and can be picked up by other players.

This is an interesting rule created to regulate and discourage reckless and violent behavior. In Second Life, some world builders can choose to make their virtual real estate a "weapon free zone".

Avatars don't go on "autopilot" when their real life personas are logged out of OASIS. They remain visible to other avatars for a certain amount of time, then fade out of the system until the user logs in again and picks up from where the avatar was last located. Although against the rules of the system, headsets can be hacked to allow multiple real life users to take over a single avatar, allowing multiple people to take turns operating one avatar.

Rules like this are created by the world builders. The hacking seems eminently possible. I wonder how the rules will be replicated within HTC's coming VR version of RPO, which aspires to re-create the OASIS.

- Samuel Steinberger and Charlie Fink

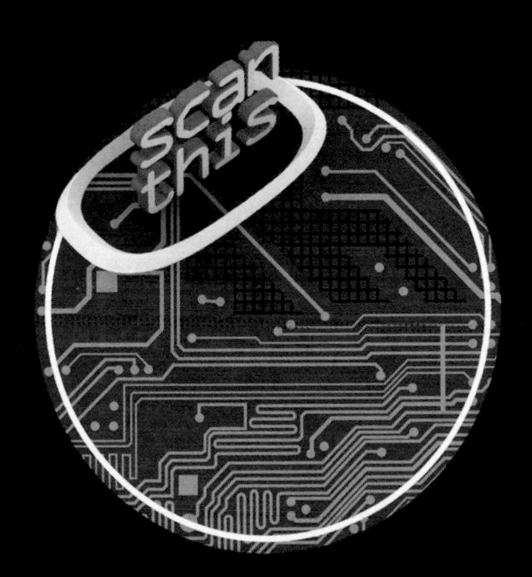

A lot of what's happening in VR and AR happened **in a dream first.**

In a fictional story

technology, often in primitive situations, such as the holographic display of Princess Leia's message for Obi-Wan Kenobi. Such a projection can be done today. Of course, you would need a projection surface. Oh, and a stereoscopic camera to do the volumetric capture. By the way, did you ever wonder who shot the Leia hologram? It's not from R2D2's perspective.
In the first *Star Wars* movie, we see the characters relieve the boredom of travel with "HoloChess", of which there are now two excellent versions. *Star Wars Jedi Challenge AR*, which I highly recommend, delivers a satisfying version of HoloChess with its new Lenovo mobile headset. Phil Tippet also recently released the new mobile game *Hologrid Monster AR*.

There is a famous scene in *Star Wars: Episode I – The Phantom Menace* (1999) in which the meeting of the Jedi Council includes members who were in distant locations by volumetrically presenting them in the same physical space as the meeting. This technology is coming to market now. Of course, an AR HMD is required.

The Terminator

The Terminator (1984) is a robot controlled by extraordinary artificial intelligence. This AI can do everything we hope to someday with an AR headset. It does real time facial and object recognition. It is contextually and geospacially aware. It does translation and displays branching dialog choice. And it learns. It adapts. It acquires and terminates targets. In short, you do not want this thing chasing you. The Terminator is sent back in time to track down Sarah Connor before she can give birth to John Connor, who will lead the rebellion against the robot masters.

When the Terminator first arrives in the past, he finds his way to a biker bar. We see from his perspective what appears to be an augmented reality display. The display scans the environment in search for tools, displaying information on vehicle models as well as environmental computations and statistics. With a quick glance around the room, the AR display sizes up a few humans, finding one of appropriate size. Upon picking this human for his body size, the terminator kindly asks for his clothes and motorcycle, which the AR display knows the man has via his boots. The man, a fan of the clothes on his back, responds defensively. After a decisive barfight, the Terminator walks out fully clothed, hopping on the correct bike with the keys he just obtained. Thanks to the initial environmental scan when he walked in, he knows which bike the keys belong to. That is one vision of what AR and AI can do together.

Minority Report

Minority Report (2002) received praise for its depiction of an augmented reality interface and hand tracking gloves. The main character can be seen manipulating images by swiping, flicking, and grabbing digital elements on a see through display. Later we see a pseudo 3D projection of the character's deceased wife projected into his living room. The image lacks detail when viewed from any direction but straight on, similar to 3D images captured by today's smartphone cameras. At another point in the film, the main character walks into what we might call a VR Arcade. A hallway filled with pods on either side lets users enjoy any fantasy imaginable.

Each of the technologies presented in *Minority Report* have influenced companies today. Touch screen displays and multi-touch software mirror the hand motions of the Precog displays. Companies like HaptX have created gloves capable of tracking a hands movement, and technologies like Leap Motion allow for hand gestures in three dimensional space, no gloves required. 3D images can be viewed with a VR head mounted display and taken with affordable cameras and smartphones. VR arcades are popping up in cities across the world, and entire VR theme parks are being built such as the $1.5B Oriental Science Fiction Park in China.

Speaking of VR theme parks...

Westworld (2016)

Westworld, based on Michael Crichton's 1973 film, is set in a wild west populated by AI driven cyborgs, programmed to provide adventure, pleasure and generally make their guests (who pay $50,000 a day for the experience) feel like they are in another world. VR headset not required. Of course, the humans misbehave badly, which causes the cyborgs to rebel. Needless to say, it doesn't end well for either the humans or the cyborgs.

The technology is not even close to being invented yet. But of course, in the distant future beyond our lives, it seems feasable. In *Westworld*, the cyborgs, including dogs, chickens, and horses, are 3D printed, and programmed with complex back stories and narratives. The cyborgs begin to dream and gain self-awareness. Made in our image, like God made Adam, our creations turn out, like Frankenstein, to be very human indeed.

I find the idea that we can't tell who is human and who is AI very relevant to the conversation about where VR is going. Right now, AI inside virtual worlds is robotic and ridiculous, and couldn't begin to pilot an avatar. But when they can, which will be soon(ish), AI will be integrated into all facets of our lives, not just video games and VR experiences.

The point is that invention is part inspiration, imagination and artistic vision, and parts hard science. But the scientists' ideas begin some place. That place is where we all begin our lives. In front of a television. Ideas, vision, use cases and pure imagination precede invention. Stories give us a way of talking about things that have not yet been invented. A lot of what's happening in VR and AR happened in a dream first. In a fictional story. The writer had to imagine a seamless use case. Every piece of art, story, movie and book we referenced is an example of this. ■

photo credit: HBO

Artificially intelligent robots, capable of desire, dreams, love, lust and merciless revenge, are the most human characters in HBO's *Westworld.*

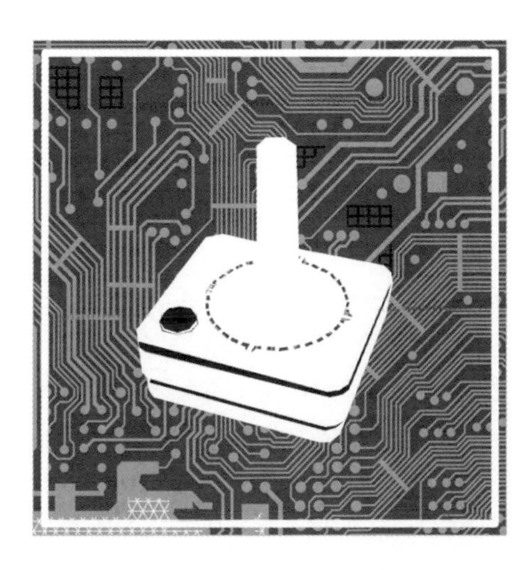

The Tech Side: HOW VR & AR

By Michael Eichenseer

The Greatest Exhibition

She marveled at the contraption. Through two eye spaced holes, she glimpsed a three dimensional world, frozen in time. So impressed, she gave the device her royal seal of approval. What was this magical device? A stereoscope designed by inventor David Brewster, creator of the kaleidoscope.

The year was 1851 and Queen Victoria had just peered through one of Brewster's early stereoscopes at the Great Exhibition. Within 5 years of Victoria's first impression, the stereoscope had sold more than half a million units and made its way to America. American Oliver Wendell Holmes produced a simplified *Holmes Stereoscope* in 1861. The same stereoscope inspired the *View Master* patent in 1939 and a generation of stereoscopes.

The first stereoscope, created by Charles Wheatstone, consisted of two mirrors angled to split a user's vision and reflect two similar images into the viewer's eyes. The images' only difference was an offset perspective from left to right. When looking into a stereoscope, each eye sees a slightly different image. The images combine and appear as a single three dimensional image. This created the illusion of three dimensions on a flat image.

David Brewster improved the design using prisms to split the viewer's vision between the stereo images, and Oliver Holmes simplified the design with two lenses spaced eye width apart with an opaque divider in between. Behind each lense was one of two near identical images, offset just enough to create a sense of depth.

WORK

The humble Holmes Stereoscope, bringing 3D images to the 19th century.

Stereopsis

A stereoscopic image is the projection of two flat images offset and overlayed on top of one another in order to produce the illusion of three dimensions. Humans see the world through stereoscopic vision, or stereopsis. Our minds see the images from both the left and right eye simultaneously, producing the three dimensions that allow us to judge distance and explore the world around us.

VR headsets today use the same technique as Brewster's stereoscope. One big difference, however, is that flat pictures have been replaced by small organic light emitting diode (OLED) screens and whatever 3D graphics a computer or smartphone is cranking out. Viewing two images through focus adjusting lenses produces a sense of depth. Users experience the same three dimensional vision they are accustomed to in the real world, but it's produced via computer graphics.

While there are many parts that make up the whole of a functioning 3D full 360 degree VR or AR experience, the main requirements can be broken down into three pieces: Optics, tracking, and haptics. These technologies have existed for decades, but it has taken vast improvements in processing, digital displays, and software to produce the VR we know today.

So, where are we at now? What evolutions have transformed the humble stereoscope of Victorian England into the latest VR rig, with its stereoscopic imagery, six degrees of freedom (6DOF) tracking and haptic feedback simulating full body interactions within a three dimensional environment? At the very least, the language has become more complicated, so let's break it down.

Optics: Latency, Resolution, and Field of View

Our eyes contain 70 percent of all of our body's sensory receptors, and 40 percent of our neocortex may be involved in visual processing. It is through vision that we largely understand and navigate the world. VR headsets rely on the power of visual perception to fool the wearers' minds and induce immersion. Queen Victoria saw a frozen image of the past. Today we see living breathing virtual worlds. To see these worlds clearly and feel immersion, we need good optics.

Optics is the study of sight and behavior of light. In the case of VR headsets, optics refers to the technologies used to produce immersive visuals for users. Lenses, displays, and graphics processing all come together to produce the optics of a VR headset.

In a VR headset you will see two lenses that bend and focus light. The lenses focus the light at an infinite focal point, instead of right next to your eyes. One display for each eye creates a stereoscopic effect. The outcome is a virtual world in crystal clear 3D.

The lenses used in the HTC Vive and Oculus Rift are called Fresnel and Hybrid Fresnel respectively. Fresnel lenses have concentric ridges which can be seen when peering into a VR headset or at the top of your favorite lighthouse. The lenses were originally designed to improve the distance that light from a lighthouse could travel. Our stereoscope designer David Brewster convinced the United Kingdom to adopt these lenses in their lighthouses back in the 1800s. Fresnel lenses can be made thinner than standard lenses and lend themselves well in the compact form factor of a VR headset.

Behind each lens is an OLED display, similar or the same as the display found on your smartphone. Smartphone, and now VR HMD screens, use a specific type of OLED known as active matrix organic light emitting diode or AMOLED, which provides faster refresh rates than standard OLED displays.

Both Vive and Rift have resolutions of 1200 × 1080 pixels per eye, or an equivalent of ~1K resolution. While the resolution is high enough to create a sense of immersion, and display a reasonable amount of information, today's displays leave much to be desired in terms of resolution.

The Screen Door Effect

The low resolution of today's displays cause a screen door effect. Everything in VR appears as if seen through a screen door. Some headsets sport displays with 4K or more pixel densities. Unfortunately, most consumer graphics cards are not built to handle such high resolutions.

Despite the screen door effect, users still feel a sense of immersion when they wear VR headsets. Because whether the resolution is low or high, the current generation of VR headsets have a high enough pixel density to warrant a flood of new VR experiences. Particularly fast paced games requiring quick reflexes and an ability to spot objectives near and far.

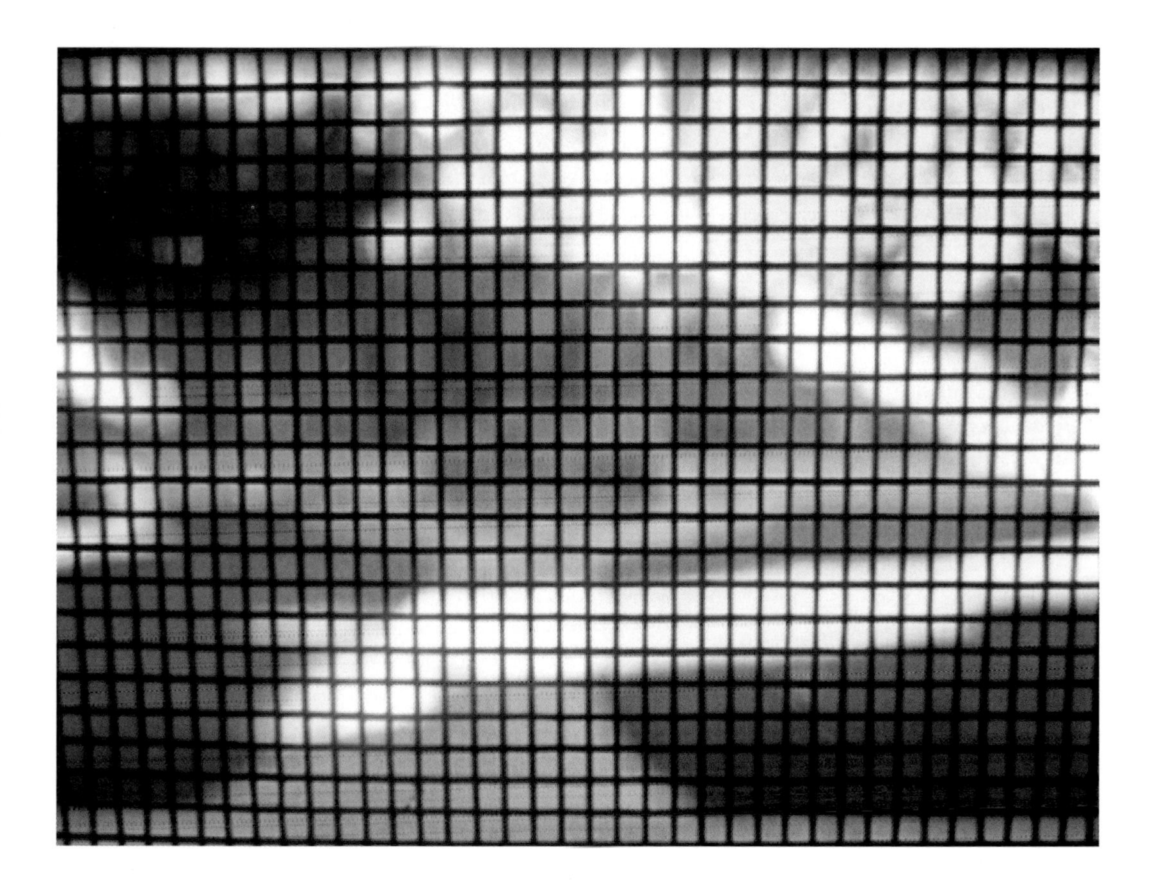

It's Time to Focus

Today's VR optics lack accommodation and depth of field. That is, there is no adjustment of focus between various depth planes of the image. The offset between the eyes simulates three dimensions, but the visual distance of the display is focused at infinity. Whether or not a user looks near or far away, the image will remain at the same clarity as if it were far away, similar to a camera with its aperture set wide open.

Groups such as Oculus Research are solving the depth of field problem with a Spatial Light Modulator (SLM), or what they call focal surface display. A spatial light modulator splits an image up into various depths, providing the multiple levels of focus our eyes are accustomed to in reality.

As the resolution and fidelity of optics improves, so does the graphics processing required. As a way to get ahead of the curve, VR developers use foveated rendering to reduce the required processing power to produce a high resolution image.

Foveated rendering uses an aspect of human sight known as foveal vision. That is, besides a small area at the center of our visual focus known as the fovea, the rest of our vision is quite blurry.

A high resolution VR image can be displayed at full resolution at the center of a user's gaze, and in reduced quality toward the edges of the visual field. Older graphics hardware could feasibly run a quality graphics display using foveated rendering to reduce processing required.

In order for foveated rendering to work, a headset must be capable of user eye tracking. As of this writing, eye tracking is native to only one headset: FOVE. Other manufacturers have stated their desire to implement eye tracking in future iterations, but so far have come up short. Foveated rendering is one of many applications for eye tracking technology, which also includes user analytics, user interface (UI) interactions, and immersive storytelling.

VR headsets rely on the power of visual perception to fool the wearers' minds and induce immersion.

Foveated rendering reduces the image quality in a user's periphery in order to save on graphics processing. Peripheral vision, though lower in quality, is a large piece of the immersion puzzle for VR headsets.

See The Big Picture

The approximate field of view in both the Oculus and the Vive is 110 degrees. In the real world, our full field of view is just over 210 degrees, so manufacturers are racing to achieve a 200-degree field of view to cover a user's entire field of vision. Those who have experienced VR gameplay can attest to the binocular effect of VR goggles. Though today's field of view cuts off much of our peripheral vision and feels similar to looking through a pair of binoculars, Queen Victoria would be impressed. Her stereoscope would feel like a peephole in comparison.

Companies like PIMAX are creating wide field of view headsets that immerse a user's entire field of vision. Increasing the field of view opens up opportunities for peripheral vision based user interfaces, along with immersive storytelling applications.

While the wide field of view present in today's VR headsets provides a compelling amount of immersion, it can also result in VR sickness. Since the image is taking up a majority of a user's field of view, the ocular nerve will attempt to sync up with what the inner ear feels. If the immersive image lags behind or moves independent of a user, dizziness and nausea can develop.

The AMOLED displays used in VR headsets today are 90Hz, meaning they refresh 90 times every second. This refresh rate is the limit for maximum frames per second (FPS) an image can be displayed. Unlike traditional television, at 24 frames per second, HD television at 60 frames per second, or traditional gaming, 30-60 fps, VR games run at 90 FPS or more in order to avoid disorientation in users.

If VR users experience lag and choppiness in their visual field they will lose immersion and may succumb to motion sickness. The image the user sees must align to their brain's perceived location in the virtual world. When their head moves, the image must move fast enough to be indistinguishable from a real world movement.

Tracking: Degrees of Freedom, Line of Sight, and Gloves

Without the ability to track a user's head position while inside VR, we would be left with an experience akin to the View Master toys of the past. While clicking through a few images can provide some entertainment, the sense of immersion is limited.

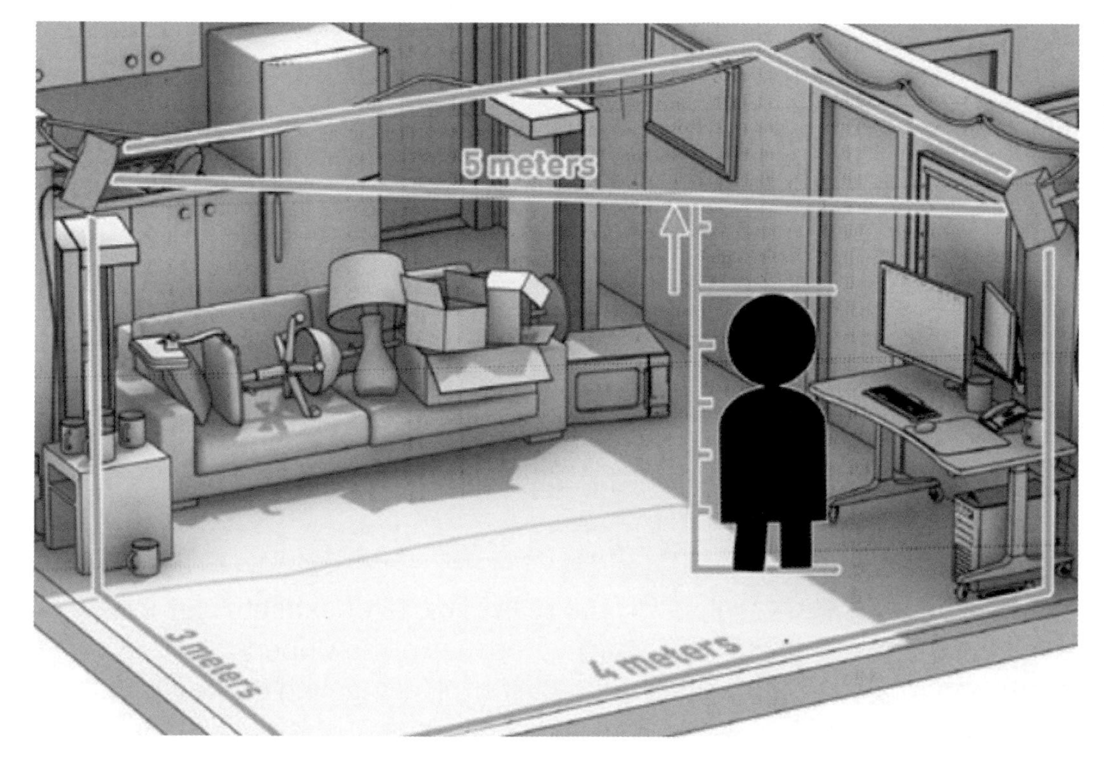

The HTC Vive "room scale" setup. Add more lighthouses, get more room. Warehouse scale "free roam" VR can use dozens.

Lego guy confirms high end VR is not yet plug and play.

The position of the user's head, or at least the direction they are facing, is tracked in order for a user to feel immersion. When a user moves his or her head, the display updates according to the new head position. These updates occur quickly, and simulate real world movement.

The time it takes for a display to update to the user's head position is called latency. Latency is a measurement of delay, usually shown in the form of milliseconds (ms), or 1/1000th of a second. Consumer level VR displays must have a latency of 15ms or less in order to avoid VR sickness in most users. Even with good frame rates, if the image lags behind the user's actual head position, VR sickness can result. Software and hardware are needed to reduce the latency to under 7ms, the lowest humans can detect.

VR sickness was a large reason VR headsets, in production since the 1990's, lagged in popularity. The hardware and software of the past were unable to produce a fast enough image to fool the human vestibular system. Thankfully, Oculus solved this problem back in 2012 by reducing latency to under 15ms, and we have since seen a plethora of high refresh rate and low latency displays.

6DOF or Just Three

There are 6 total degrees of freedom when it comes to movement in three dimensional space: Surging, swaying, heaving, rolling, pitching, and yawing.

Surging *is when a user moves his or her head forward and backward, such as leaning toward an object for closer inspection.*

Swaying *is side to side movement, such as leaning around a corner.*

Heaving *is the up and down movement of the head, crouching and jumping.*

The latter three are more traditional indicators, all having to do with the rotation of a user's head within three dimensional space: Turning, tilting, and nodding. And are known together as three degrees of freedom, or 3DOF.

Rolling *is tilting side to side, such as a cute curious puppy dog with perked up ears.*

Pitching *is the nodding of a user's head, looking up or down.*

Yawing *is rotating left and right, turning around to look behind you.*

3DOF is used for smartphone based virtual reality, such as the Google Daydream. The rotation of a user's head is tracked by the gyroscope inside the smartphone. Users can look left and right, up and down, and tilt their heads. If a user leans forward or backward, their position within the virtual environment will not be updated. This lack of positional tracking has been known to cause motion sickness in users when the visuals they see do not match up with the movement they feel.

Useful for viewing 360 photos or videos, a lack of positional tracking leaves much to be desired in the realm of immersion within VR. It is for this reason the Rift and Vive have developed 6DOF tracking technology for their headsets. This means not only a user's view direction, but their head's position in 3D space is tracked and translated into VR. When you lean to the right, you move within the virtual world.

To achieve 6DOF tracking requires one of two techniques: Inside out tracking, or outside in tracking. Both of which supply VR software with the positional data required to update a user's head position within VR. Most VR headsets today rely on outside in tracking, requiring users to set up external sensors surrounding their VR use space.

Outside In Tracking

With outside in tracking, each object being tracked can be seen by a camera or lighthouse. Its position, both rotation and translation, can be determined via software connected to the cameras. In addition, both the Oculus and Vive use a gyroscope to determine the rotation of a user's head.

If you've ever seen a motion capture session where actors are fitted with a series of white balls, you'll have an idea of a Rift camera setup. Instead of white balls attached to a motion capture suit, small infrared lights are attached to the headset itself. The Rift uses two or more cameras to track the

Nolan Bushnell said at ARinAction at MIT in early 2017 that AR is in the "Pong" stage of development. Indeed, industry conferences are dominated not by the makers of consumer products, but by the companies that make the tools that enable consumer software products to be made. No one can have a conversation about developing content for VR and AR without talking about Unity.

Unity makes the software tools used by more than four million game developers. Because of its unique 3D qualities, Unity is quickly becoming one of the formative Metaverse's most important, and most valuable, companies. Unity's suite of tools, and its asset store, where developers can buy designs and pre-coded actions, are simple enough for a most designers to work with. No coding skills required. Unity just raised an eye popping $400 million from private equity firm Silver Lake, raising the value of the private company to $2.8 billion. Unity Chief Executive John Riccitiello told Bloomberg there are "a lot of investors and investor interest is the fact that we have about 70 percent of AR and VR content built on Unity." Unity is going to become more valuable to more people because in a virtual world, as users seek to acquire 3D objects and tools they themselves can utilize in the virtual world of their choice. Some of the most valuable companies in the world, like Oracle (market cap $242 billion), are little known to the public, but provide the connective tissue that makes the services we depend on, like electronic banking, work.

Film schools not teaching their students how to build worlds and make movies with Unity are committing malpractice.

position of infrared LEDs located around the headset and touch controllers. The cameras can see infrared light in the formation necessary for software to read and calculate a user's position. The Vive uses a different system to handle outside in tracking. Two lighthouses, boxes filled with LEDs and laser emitters, sweep a series of lasers across a play area. The Vive headset and controllers are outfitted with multiple sensors capable of detecting the swath of lasers from the devices. By calculating the timing of each sensor being hit by a laser, software can track a user's three dimensional position.

In both cases, a line of sight is required between the tracked object, in this case a user's headset, and the camera/lighthouse. If line of sight is lost, the tracking of the user in VR will also be lost. Users of VR might experience this when reaching a hand outside of their play area. The hand loses tracking, causing the user's virtual hand to drift into space, away from the user's physical position. If this happens to the headset, and not just a user's virtual hand, he or she can become disoriented and dizzy.

Adam is a Webby Award-winning short film created with the Unity game engine and rendered in real time. It was built to showcase and test out the graphical quality achievable with Unity in 2016. The well received short will be expanded by acclaimed science film director Neill Blomkamp (District 9) into a science fiction franchise *ADAM: The Mirror* and *ADAM: The Prophet*.

Created and produced by Blomkamp's OATS Studios using the latest version of Unity 2017, these new short films will showcase the power of working within an integrated real time environment, empowering the team to build, texture, animate, light, and render all in Unity to deliver high quality graphics at a fraction of the cost and time of a normal film production cycle.

There is something very important going on with Unity and Hollywood that is going to impact media production dramatically in the coming years as game production techniques begin to supplant hundred year old narrative entertainment production methods. This is going to deeply disrupt the film and television industry as we know it. Filmmakers, cinematographers, designers and technicians have a great opportunity to harness disruption to their advantage.

Young people entering the business today are uniquely positioned to ride this wave of disruption into new leadership roles in content creation. Skip that production assistant grind. Compete with the studio by building your own movie with Unity. There are already hundreds of high paying jobs in game development and film production for skilled Unity developers listed on Linkedin as I write this. Anyone can learn it. There are numerous free and paid online courses, and even University and Community College extension courses about Unity now taught across the country. Film schools not teaching their students how to build worlds and make movies with Unity are committing malpractice.

Inside Out Tracking

Inside out tracking looks out at the world in order to determine position in three dimensional space. Cameras on the headset record the outside world and determine the position of the headset by comparing it to the static positions of real world objects. Inside out tracking does not require external sensors in order to work. Inside out tracking does require sophisticated software in order to function well, software similar to Apple's ARkit or Google's ARCore, both of which use a smartphone camera to calculate depth and location in three-dimensional space. The same technology used to place lovable Pokemon onto the real world can be used to position a user's head in the virtual world.

Inside out tracking is seen as a requirement for future untethered VR devices. It makes little sense to place sensors throughout our environment when the VR devices themselves could track their positions.

Head and Shoulders, Knees and 'Bows

The most important piece of hardware to be tracked in VR is the headset. If a user's head position is not well tracked, there will be a loss of immersion, or worse, VR sickness. When the visuals of VR do not align with a user's inner ear, nausea is likely to develop. With today's VR headsets, most users can no longer detect any delay between what they see and what they feel.

Hands are the next piece of the VR tracking picture. The big VR hardware manufacturers such as HTC, Oculus, and Microsoft partners today all provide a means to track the position of a user's hands within VR. Oculus Rift has Touch controllers and the HTC Vive has wands. Since humans use their hands for a majority of their interactions with the real world, it makes sense to bring the hands into VR as well. When a user can reach out and touch a virtual object, the level of immersion rises.

Today's hand controllers rely on the same outside in tracking as their headset counterparts. There are some VR hand controllers, such as the Google Daydream's wand, that are tracked by a camera located on the headset. The downside to this approach is the need for line of sight between the headset and the controller. A user must hold the controller out in front of them where the camera can see it. Whereas room scale systems like the Rift and Vive detect the location of the hand controllers separate from the position of the headset. A room scale VR user, for example, can reach behind his or her head to unsheath a sword.

For those who don't wish to hold a controller, Leap Motion uses infrared lights and two cameras to track the movements of a user's hands. While this provides no haptic feedback, it allows for high fidelity interactions using all ten fingers. The issue again is line of sight. If a hand or single finger is bent away from the Leap Motion sensor, its position can no longer be tracked.

Beyond the hands, the Microsoft Kinect uses optical depth sensing technology to achieve full body tracking. Since the days of the Xbox 360, developers have used the Kinect to achieve full body tracking within virtual environments. Developers still use the Kinect today when testing applications for virtual reality, although it suffers from the same line of sight issue. Once a user's limbs can't be seen by the two Kinect cameras, tracking is lost.

The leading VR systems in 2017 do not track a user's full body. Only the head and hands make their way into VR. The arms, torso, and legs must all be generated through software, or not at all. Many games, such as Rec Room choose to leave out a user's legs altogether.

Full body tracking is possible with today's technology, however, and companies are racing to create a consumer level solution. HTC released Vive pucks

HaptX

How we interact with the virtual worlds we can now occupy is one of the central questions of VR.

When I first saw this iconic VR full immersion rig, I was intrigued by both the vision and the assumptions behind it. Of course, tens of thousands of developers would have to incorporate this haptic rig via a Software Development Kit (SDK) in order for the product to be viable. It just didn't make a lot of sense. Then I read Ernest Cline's seminal novel, Ready Player One and realized that it's not so far-fetched. Everyone wants to be seamlessly, effortlessly and completely immersed in virtual reality. But there's no consensus on how to address it, or ultimately how important it is.

How we interact with the virtual worlds we can now occupy is one of the central questions of VR. The interface will not be the same for many of the things we do now, from the basics of locomotion (teleporting vs. walking) to complex tasks like interacting with a workstation. Haptic gloves are part of the basic rig in Ready Player One. When the characters play 1980s arcade games inside VR, they feel buttons and joysticks. On a practical side, haptical gloves would allow surgeons to feel a scalpel in their hands, and get realistic feedback when using it. For VR flight simulators to supplant the expensive physical cockpits used to train airline pilots, they'd need to provide accurate feedback from a wide array inputs, including switches and touch screens, to say nothing of gripping the yoke at takeoff and landing.

Jake Rubin, Co-Founder and CEO of HaptX, spent half his life envisioning full body haptic system. When researching how to build this ambitious project, he found Dr. Bob Crockett, Chair of the Biomedical Engineering Department at California Polytechnic State University. Rubin called Dr. Crockett, and after weeks of digging into one another's research, they founded HaptX. Since then, the company has raised over $9 million in funding, and announced its first product: HaptX Gloves.

HaptX has decided to focus on haptic gloves because they can be made at a reasonable price, at least for enterprise customers, and there is demand from actual customers who have seen the HaptX prototype. In the next several years, the company expects to reduce the cost to a place where gloves become viable for consumers.

WHEN A USER CAN **reach out and touch** A VIRTUAL OBJECT, **the level of immersion rises** in 2017 as a means to track objects other than the headset or controllers. The pucks have been placed on shoes and belts in order to track the feet and waists of users, simulating full body tracking. Software can extrapolate the appropriate location of the remainder of a user's limbs. IKinema's project Orion produced a compelling demo showing full body movement replicated in VR with only feet, hands, waist, and head being tracked.

Haptics: It's Easier To Press A Button in VR If You Can Feel It

In some instances, the hardware built for tracking a user's body can double as an interface for haptic feedback. For example, the touch controllers for the Oculus Rift contain rumble packs to provide physical feedback to a user's hands. The controllers can vibrate when a user interacts with virtual objects. The applications are limited, but useful for increasing immersion.

The future is in full hand tracking and haptic feedback. Full hand haptics allow designers to simulate controllers, keyboards, objects, etc. of any size and shape within virtual reality. Instead of hardware companies competing to provide a plethora of VR hardware, software designers can build interfaces within VR utilizing the haptic feedback of a full featured VR glove.

VR gloves have been popularized by science fiction stories like *Snow Crash*, *Minority Report*, and *Ready Player One*. Each finger's position is tracked by the hardware of the glove, and the forces applied to each finger can be adjusted. By increasing the pressure required to bend a finger of the glove, a designer can simulate the feeling of pressing a finger against something in VR. The hilt of a virtual sword might stop a user from fully closing their hand. In reality, the user is holding their hand in a semi-closed fist, while in VR they grasp a sword.

Facebook has stated their intention to develop full featured VR gloves for the Rift platform. Current iterations of VR controllers, like Valve's "Knuckle" controller and Rift Touch use capacitive sensors, which detect nearby conductive human fingers and their approximate distance from the controller. Startups, such as VRGluv, are developing VR gloves featuring full tracking and haptic force feedback.

Some forms of haptic feedback don't require users to wear gloves. Ultrahaptics uses sound waves to simulate buttons and dials a user can reach out and feel. An ultrasonic wave pulses just enough air towards a user's fingers to simulate the feeling of touch. So far this technology has tested well for simple user interfaces. I.e. Temperature dials on a kitchen stove. Development continues for more complex interfaces such as full body VR gaming.

Full Body Haptics

The ultimate immersive VR experience will require full body haptic feedback. Not only will a user's body appear in VR in real time, but they will feel the virtual world through physical touch and force feedback. Player's bodies have appeared in the virtual world since the Xbox Kinect, and soon, haptic feedback will let them feel that world.

By placing rumble packs on key points of the body, the HardlightVR suit allows users to "feel" their VR environment. The main application being games where projectiles can hit a player. By recording the location of a virtual bullet hit, the haptic suit can relay that location and vibrate accordingly. HardlightVR ran a successful kickstarter campaign for their VR haptic suit in early 2017.

Teslasuit is developing a similar full body haptic suit. The suit contains strategically placed rumble packs, motion capture sensors, and a capillary based climate control system. The current iteration of the Teslasuit wears like a jumpsuit and has yet to make its way into the homes of consumers. But the ability to provide both physical and temperature based feedback gives a tantalizing look at the future of VR immersion.

HaptX uses mechanical arms and actuators attached to each leg to simulate the haptic feedback from walking, running, and climbing stairs. Currently built for enterprise applications, actuating arms may be the future of full body VR rigs. While omni-directional treadmills make for a simple VR locomotion solution, mechanical arms provide realistic force feedback for a user's full body.

Haptic feedback increases the real world applications of VR. Surgeons today practice precise incisions with virtual scalpels designed for realistic physical feedback. A mechanical arm attached to the handle of a virtual scalpel simulates the physical feedback a surgeon feels while making an incision. Military personnel, police, firefighters, and all manner of dangerous professions can use full body haptic feedback, from systems like HaptX, to safely train for dangerous real world situations.

Get It While It's Hot!

Temperature is another form of haptic feedback, and another way to increase immersion. Large installations such as The VOID do their best to maintain an appropriate temperature for the VR experience using standard climate control systems any large building uses. Personal systems like the Teslasuit rely on a capillary system to move temperature controlled liquid around the user's body. Teslasuit has their own HaptX system capable of simulating warm and cold temperatures in real time on a surface at least as large as a human hand.

A common complaint of VR users today is sweat. Having a headset strapped

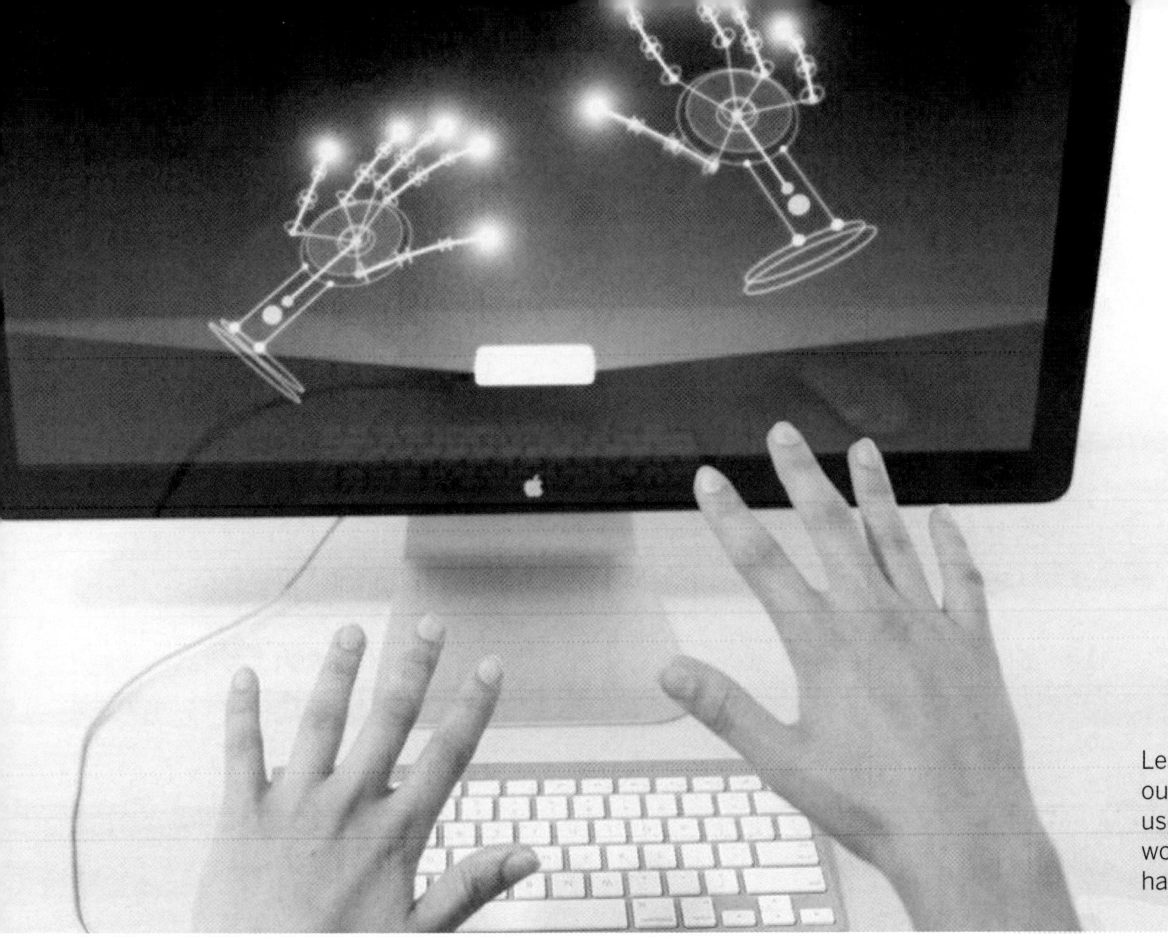

Leap Motion has an inside out facing tracker that puts users' hands into the virtual world, potentially eliminating hand controllers.

photo credit: Leap Motion

to your face while you swing your arms around can work up a sweat even in the fittest of users. One product, Vive N Chill, attaches small fans to the user's headset in order to blow air across a user's face. But what about using airflow for immersion?

Some companies see wind as a much needed addition to an immersive VR experience. Birdly, a VR flight simulator built by SOMNIACS, may not be the flight simulator of you imagine. It places users face down with arms stretched wide. By physically flapping one's arms, the user experiences a sense of flight mimicking that of a bird. On the front of the Birdly flight simulator is a fan tied to the VR experience itself, blowing harder or softer depending on the user's virtual speed. For the most part, airflow has a long way to go before hitting consumer level devices.

Do You Hear That? The Future Is Calling

Three dimensional audio is easy to overlook when discussing cutting edge virtual reality hardware, but immensely important to immersion. Just think of how important directional sound is in gaming systems, and home and movie theaters. Gamers have enjoyed the hard work of sound engineers in first person shooters, helping players follow the action. Movies and home theaters use speakers throughout the screening room to immerse viewers in the film. Even musicians create movement by mixing their songs to play between the left and right speakers.

Virtual reality and its ability to track user position in three dimensional space makes positional audio a no-brainer. The technology behind 3D audio has been developing steadily since the 90's. Alongside PC and console gaming and the Oculus Rift and HTC Vive headsets with headphones attached, are the latest sound producing iterations, ready to provide 3D audio effects using standard stereo speakers or headphones.

Our brains are wired to locate sounds in 3D space using only our two ears as inputs. Thanks to head related transfer functions (HRTF), a cheap pair of stereo headphones can simulate sound sources in three dimensional space. By modifying reverberations, sound levels, and other technical aspects of sound, HRTF can simulate 3D audio with high accuracy. HRTF modifies stereo sound to mimic the real- world effects of distance and location between sound sources and our ears. Coupled with positional tracking and a stereo headset, HRTF provides an immersive VR audio experience without unnecessary hardware complexity.

In the future we may see more complex 3D audio hardware integrated into VR systems, but utilizing HRTF and stereo speakers represents a large cost savings today.

So Real You Can Taste It

On the fringe of VR development lies the world of taste and smell, two closely related senses not immediately considered when imagining virtual reality. Many companies aim to change our relationship with food using virtual reality. Not only can they simulate the visuals of your favorite dishes, they can simulate the way your food smells, tastes, and chews.

South Park creators had fun with the concept of smelling the virtual world via their Nosulus Rift, a peripheral for their non-VR game *The Fractured But Whole*. While created as a joke, the technology behind the device may make its way into VR systems everywhere.

Project Nourished has brought more than sight and touch to VR. They have brought tastes, smells, and textures. And not just the textures we feel with our hands, but the texture of our food. With a mix of electrodes, atomizers, and neutral tasting gelatin, Project Nourished can simulate the taste and texture of real, and imagined, foods within virtual reality. After an atomizer releases an aroma, electrodes stimulate the jaw muscles alongside visual and audio queues to create a physical sense of chewing, tasting, and smelling specific foods.

It is feasible to simulate a plethora of smells using a limited number of chemicals. These chemicals can be located in small vials attached to a user's VR headset or in an atomizer nearby. When software is triggered by a user in VR, a light mist is sprayed toward the user's nose. Startups like Vaqso are betting that smelling your virtual environment adds another layer of immersion.

Taste is a bit trickier than smell. Japanese researcher Takuji Narumi's olfactory VR headset has fooled subjects' senses by making a plain cookie taste like something different, such as a lemon cookie via a mix of aromatics and visual stimuli.

Nimesha Ranasinghe has successfully produced the flavors of sour and salt using an electrode. Dubbed a "digital taste interface" or "digital lolipop", an electrode produces electrical currents on a user's tongue in order to simulate the sensations of taste. Coupled with aromatics, designers can simulate any number of tastes within VR to great effect.

Looking Forward

Technology enabling users to feel immersed in a virtual world with all five senses can be simulated to an impressive degree today. Startups, research labs, and enthusiasts are clamouring to create true immersion in the virtual world, not just in a lab. Deep immersion will soon reach the living rooms and offices of users worldwide. The future is already here. It won't be long before the VR of our dreams arrives on our doorstep. ■

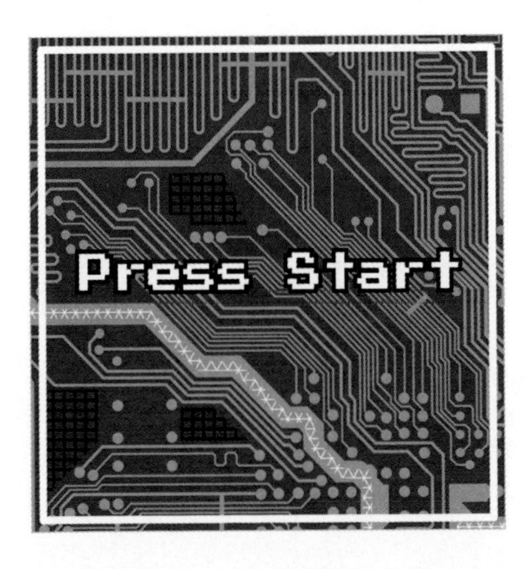

VR & AR BUYERS GUIDE

FROM $14.99 TO $599

By Charlie Fink, Sam Steinberger, Michael Eichenseer & Samantha Wolfe

With the arrival of Windows MR compatible headsets and a new consumer augmented reality HMD for mobile Star Wars games, there are plenty of options for those interested in entering the world of VR and AR. Buyers will find is there is a robust and growing selection of both free and paid experiences now, including new VR versions of AAA games like Doom and Fallout. The bad news is that for the high end Oculus Rift and HTC Vive, a gaming computer, which features a super fast advanced graphics card, is required. These computers have not come down in price, one of the factors that forced the makers of high-end VR, including Playstation VR, to cut prices. This helped 10%. The fact that there are only 25M VR capable PCs in the market, and their stubbornly high prices, puts high-end VR out of reach for most consumers. If you already own a Playstation 4, you're only $300 away from the best VR has to offer in the home, which has made PSVR the most popular advanced platform, with nearly two million sold (out of a VR-capable install base of 60M consoles).

There are several good lower cost mobile VR options for less than $100, provided you have a compatible phone. For Google Daydream there is the well reviewed new Pixel 2 (the camera is crazy good), and for Samsung Gear VR (which uses the Oculus store) you need their Galaxy phone.

At the moment, I'm partial to the Star Wars Jedi Challenge, a very powerful mobile AR head mounted display from Lenovo for $200. If you love Star Wars and want to get a glimpse of what real AR entertainment looks like, you have to buy this.

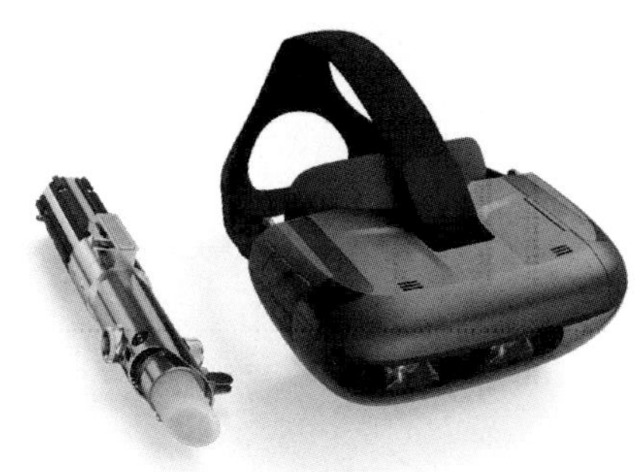

Disney

It all started with the idea of an AR light sabre in Mike Goslin's group at Disney advanced consumer products. Holochess isn't just for Wookies anymore.

Star Wars Jedi Challenge AR $199

Slide your cell phone into the top of the plastic headset, fire up your lightsaber, and fight the stereoscopic enemies beamed onto the plastic visor of the headset, creating the illusion of ghostlike "holograms" in the real world. Jedi Challenge comes with three games: Lightsaber Battles, Strategic Combat, and Holochess. While progressing through the galaxy, players can achieve multiple Jedi mastery levels from Padawan to Knight and eventually Master. As you unlock the ranks of the Jedi Order, the lightsaber will change color. A must have for Star Wars fans, it's possible with frequent updates that Star Wars AR could be with us for some time.

Acer

Acer's Windows MR headset is lightweight and features a flip up visor and microphone. It sets up in minutes.

Windows MR with The New Acer HMD $399

Was it getting dark already? I'd just taken a five minute breather from playing Steam VR's Paintball and realized I'd been playing for a solid four hours. I hadn't eaten. I'd forgotten about doing my laundry. I'd completely lost track of time. My first experience with Acer's Windows Mixed Reality headset was a smashing success. But before I get into the details, let's take a look at the device itself.

The Acer Windows Mixed Reality HMD is a thing of beauty. The black and blue visor is lightweight and comfortable, even over glasses. It's an inside-out HMD, which means the position of the visor is tracked by the visor itself, not by additional outside devices that need to be installed. It also means it's the perfect size for my tiny New York City apartment living room, which is smaller than most bedrooms.

For my first foray into Windows MR, I invited my girlfriend and another friend over to check out the headset and provide a broader sample of the experience. When we first paired the controllers with the computer there was an audible "ooh" as they lit up. The controllers are responsive. Although, after a few hours of action packed VR, I found my hands losing their grip at times and cramping up.

We followed the instructions on the screen to calibrate the set and fed the meager boundaries of my living room into the software. As the Windows Mixed Reality software launched, I was transported to an airy, spacious home on the side of a cliff, overlooking fantastical floating islands, with a backyard view of snow-covered mountains. My home theater had a retractable roof. I could walk around my house, peg photos on the walls, and watch Penrose's The Rose and I, a short VR movie.

While my friend and girlfriend took the same VR home walk around, I set up a Steam VR account. I'd heard about Rec Room's Paintball game and wanted to give it a go. Little did I realize at the time that I was about to wave goodbye to all my afternoon plans.

For the rest of the day the three of us alternated playing VR paintball, watching VR movies, and exploring what Windows MR had to offer. I wasn't disappointed, and my girlfriend said she was still seeing paintballs flying at her, even after shutting down the HMD for the night. A sure sign of a good experience. It's made me a VR believer, and judging by the number of newcomers flinging virtual paintballs at each other, I'm not alone. - Samuel Steinberger

HTC Vive

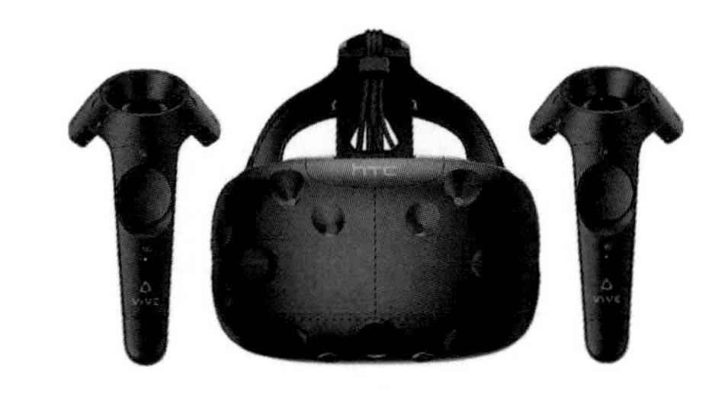

HTC Vive and controllers $599

Oculus Rift ($399), HTC Vive ($599), Playstation VR ($299)

The Rift, Vive, and PSVR are the premier gaming VR headsets in the market. Each headset has its benefits and limitations, and each one works best for certain occasions. If you're interested in VR, you can't go wrong with any of these three headsets. They are all content rich including experiences that are cross platform between all three.

The PSVR is an easy choice if your household already has a PS4 in the living room.

Within minutes your living room turns into a virtual world. However, PSVR is made for sitting or standing, not moving around. If you do desire full room scale play, you'll have to look into the PC connected VR headsets.

The Oculus Rift has made waves with its precipitous price drops. Currently tagged at $399 for the headset and touch controllers, the price makes the cost of a computer capable of running an Oculus Rift somewhat more palatable. There is an extra cost of $60 for a third camera if you want to play full room scale. After spending a thousand or more for a PC capable of running a Rift, you will be very impressed with the selection of exclusive games and the high amount of cross platform play.

Oculus

Oculus Rift and Touch Controller bundle, now $399, two hundred dollars less than the HTC Vive

The HTC Vive offers a headset designed from the ground up for room-scale with access to a massive library of Steam games. Whereas, the Rift needs a separate third camera in order to smoothly operate in room scale, the Vive does so with two lighthouses, both of which do not have to be connected via USB to the computer. Coming in at $599 alongside a VR ready PC makes the Vive the most expensive of the bunch, though not much more than a room scale ready Rift + Touch.

Sony

At $299 Sony Playstation VR is the low price leader. Sony said it has sold over 1 M units. More than Vive and Rift combined.

The PSVR may be the most successful of the VR headsets so far. The games are great too, with tons of PS4 exclusives. PSVR has one of the only AAA VR titles to date: Resident Evil 7, as well as Skyrim VR. An Aim Controller in the shape of two handed gun comes bundled with Farpoint, a PSVR exclusive shooter, and will cost you $80.

There are quite a few free experiences for the Rift that make the headset a bargain, as well as high quality exclusives developed through the Oculus developer program. One of the most popular multiplayer VR games, Echo Arena, is exclusive to the Rift at this time, pitting teams of 3 players against each other in a zero gravity ultimate frisbee style arena.

The Vive has the largest selection of titles to play, many of which are small games built by indie developers. Most games for the Vive are cross platform with the Rift, if not the PSVR too, but that does not stop Vive exclusives from popping up. A prime example is Fallout 4 VR.

Vive thinks this AAA title from Bethesda Softworks will move a lot of headsets.

Each of the three headsets are quite similar when it comes to visual quality. PSVR is the only system that is essentially plug and play. Both the Vive and Rift require base stations/cameras to be set up around the play space, and calibration to take place before a user can play.

Room scale is what the Vive was built for, and the Rift can achieve room scale with a third camera, but the PSVR lacks full room scale.

Assuming you already have a high end gaming PC, it's up to you to choose which headset you prefer. But if you have a PS4 at home, then PSVR is the way to go. A PS4 Pro is recommended for PSVR however. - Michael Eichenseer

Google

Google Daydream is the most comfortable of the mobile VR headsets.

Google Daydream (Mobile VR) $99

Comfort and flexibility aren't words you normally associate with VR. However, if you prioritize those qualities, Google's $99 Daydream View is the mobile phone headset for you. The second generation 360 degree video mobile device has three choices of color fabric (charcoal, fog, and coral), a comfortable shape, and, as long as your phone is Daydream compatible (such as the Pixel 2, Moto Z, Galaxy S8), you're good to go.

Like its cousin, Google Cardboard, you simply place your phone in the headset with no need to connect it electronically. However, if you must wear glasses with a headset, try the Daydream on before committing. Although not official, analysts have estimated that there will be about 6.8 million of Google Daydreams shipped in 2017. That's pretty impressive for the new kid on the block. The Daydream headset uses Google Play (like other Android devices) and has about 200 apps (about half for free), up from in 2016, with more on the way. The Daydream has some exclusive apps including Google's Play Movies and YouTube. The Daydream's wireless remote controller is integrated into every game allowing a certain amount of interactivity for driving a virtual car or yielding a sword. All this adds up to a comfortable headset that lets you get lost in a 360-degree world for hours at a time. - Samantha Wolfe

Samsung

The Samsung Gear has shipped nearly 10M units since its introduction in 2015, making it by far the most popular mobile VR headset (unless you count Google Cardboard).

Samsung Gear VR $129

If you're looking for a more substantial mobile phone headset with a deeper library of content, look no further than the $129 Samsung Gear VR. The 2017 Oculus-powered headset is lighter than 2016's version, has a wide 101 degree field-of-view, a wireless controller, padding that easily accommodates glasses, and a focus adjustment on top of the device. All of this adds up to a better feeling of immersion than other mobile headsets. This is why it's sold over 8 million units making it the most popular out there, besides the entry level Google Cardboard.

The Samsung Gear VR offers access to the rich library of games and experiences within the Oculus store. The set up is a bit awkward, but the content available is worth it. There's well known TV and movie title offshoots like Stranger Things: Face Your Fears and Blade Runner 2049: Replicant Pursuit, as well as favorites such as Minecraft Gear VR Edition, Eve Gunjack, and NextVR for live sports and music in 360.

As the Gear VR requires a Samsung Galaxy phone (S6 or later), if you have a Samsung phone or were considering getting one, this is the perfect headset choice. If you purchase the Samsung Galaxy S8 or the Galaxy Note8 Edition, it's packaged with the Samsung Gear VR and compatible with the Google Daydream. - Samantha Wolfe

Merge

At $14.99 the Merge Cube is the ultimate mobile AR stocking stuffer.

Merge Cube $14.99 & Merge AR/VR Goggles $49.99

The best thing about the Merge Cube is that it's only $14.99, and it'll keep your kids busy for hours, perhaps longer. The Merge Cube will work with older smartphones that cannot otherwise do AR, but it may cause them to run hot and drain batteries. The award winning soft foam Merge goggles are sold everywhere. It works with the Merge Cube and any VR apps on the phone. A best seller. - Charlie Fink

ZapBox

At $30, ZapBox makes "Magic Leap magic cheap".

ZapBox Mixed Reality $30

It's possible that with the introduction of depth sensing cameras using Apple's ARKit, marker AR may have its moment. ZapBox makes "Magic Leap magic cheap" with its unique Cardboard HMD and controllers for your mobile phone. This is some good clean fun at a great price with a couple of dozen third party apps. ■

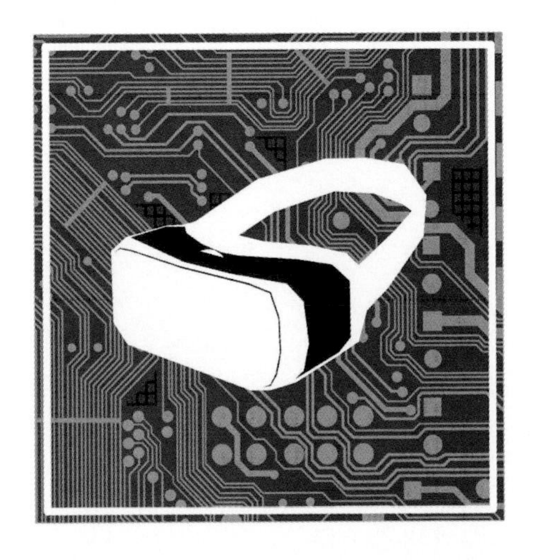

THE CONSUMER Inflection POINT

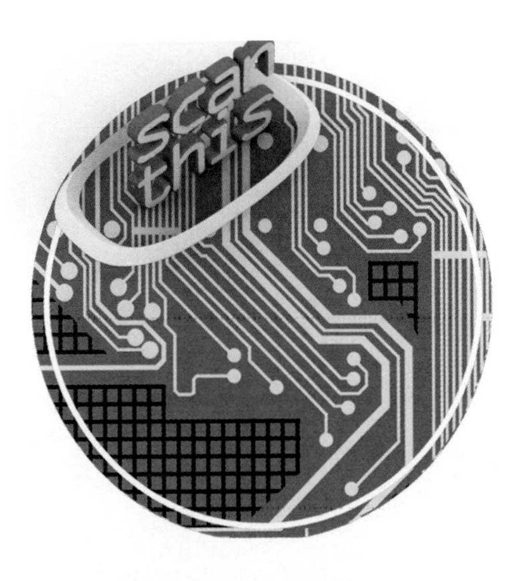
scan
this

Fashionable, low priced, [illegible]ly good for taking ten second videos for use on Snapchat, Snap Spectacles never caught on.

Why Consumer Adoption of AR & VR Will Be Slow

The transition to head worn mobile computing faces significant barriers. Unlike the smartphone, this requires big changes in consumer behavior. Head mounted displays (HMDs) are a new idea. In order to get people to buy Pepsi, they have to know what soda is. For this reason, adoption may look more like personal computers, which took fifteen years, than smartphones, which took two years.

During the Internet explosion in the early 1990s, we often looked at this graph, which shows rates of consumer technology adoption. The data suggested that the speed of adoption would continue to accelerate, which proved to be true for smartphones and tablets, but those devices took what we were already doing and made it much better.

It took fifty years to electrify the country. It took thirty years to wire landline phones. It took radio twenty years. Television, ten. The Internet took less than five years. AR and VR cannot be conflated with these technologies. Instead, it is like the personal computer, which took fifteen years to hit an inflection point. Personal computers came into our lives very slowly.

Throughout the 80s, personal computers were considered first adopter novelty items for nerds and rich people. It wasn't until the end of the decade that PCs were common in most offices. They were expensive. They ran expensive CD-ROMs, which were either games or educational in nature. If

Rapid advances in smartphones have spoiled us. VR and AR isn't going to be like that.

the computer had a modem (it was considered a peripheral, like speakers), you had to open it with a separate program. I remember in 1993 I needed to open several programs to get onto the Internet. One for TCP/IP. One for the modem itself. One for my sleek new Netscape Navigator web browser, and yet another for IRC (chat).

However, once the PC met online services, the PC hit an immediate inflection point. This happened within months. The advent of online services like AOL and Prodigy, with their all-in-one discs that brought all the disparate Internet software together into one simple (sort of) plug and play program, pushed the PC to an inflection point. By 1996, everyone had to have one, because at that point, the value proposition was so clear and substantial.

In the early 2000s, many people were given their first smartphone at work, the BlackBerry, which allowed users to send email on the go. Soon, consumer cellphones had those features, and people received remarkable upgrades for free as part of their normal cellphone replacement cycle. The wireless providers and handset makers took what we were already doing and made it much, much better. Yes, please!

Mobile AR, which turns the camera into the window through which we see the world, has been available on Android phones since 2015 and on iPhones since the fall of 2017. Because of Apple's scale, within a few days, hundreds of people could do much more with the phone. There were just two problems. The first was apps. They're novelties and game enhancements. Second, holding one's arm out to view the world through the camera may be the worst form factor accidentally invented by man.

Augmented reality works exceptionally well in enterprise (as computers did in the 80s), but they largely aren't for consumers, although there are some nifty AR-enabled toys and books (like this one!). For consumers, AR headsets are in a protean state. There are basic problems with optics and field of view. Costs are still going up, not down. Interface solutions are not obvious. Speculation swirls around the big companies and some stealthy startups (most notably Magic Leap).

Ironically, the really big utility problems are outside the smartphone. They're in the cloud, and pertain to unsolved issues of bandwidth, compression, artificial intelligence, and the lack of a geospacial social AR Cloud that would make the glasses smarter than our computers today. For VR the problems are simpler and more profound. Navigating with hand controllers is extremely awkward and can cause motion sickness. The optics are terrible. At current resolutions, the pixels are visible, creating a "screen door" effect. Even advanced headsets only have a 110 degree field of view.

Rapid advances in smartphones have spoiled us. VR and AR isn't [sic] going to be like that. ■

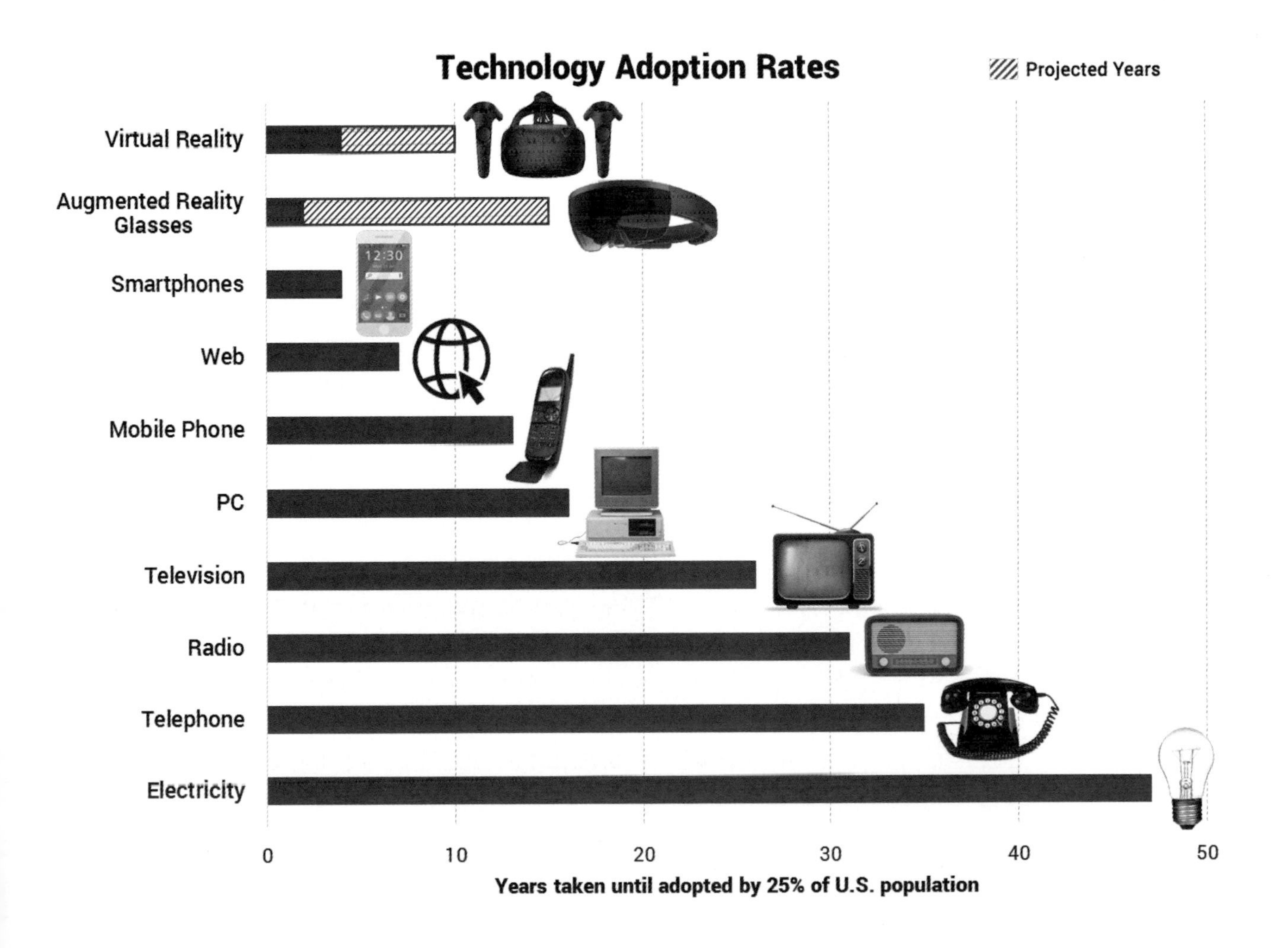

The Road To CONSUMER AR

By Peter Wilkins

Peter Wilkins is the Founder & CEO of Emergent VR. Emergent creates virtual reality for everyone. Peter created the first 3D 360 movie for consumer VR in 2014, featured at Sundance New Horizons. Before VR, he worked on Academy Award nominated films at DreamWorks Animation.

With the trough of disillusionment well underway for virtual reality (VR), our industry has turned to augmented reality (AR) as the next great hope. AR is even more technically challenging to productize than VR, but offers potentially greater rewards. What is it going to take to build a robust, vibrant AR ecosystem on par with the App Store for iPhone? We need four components to create the platform for widespread AR adoption:

- Spatial Services
- Artificial Intelligence (AI)
- User Interface (UI)
- Hardware

With these four components together in the form of a software development kit (SDK), developers can begin to build truly useful and compelling AR applications.

Spatial Services

Truly useful AR requires the device to understand your context in the world; not just a two dimensional location on the surface of Earth such as Google Maps, but a full three dimensional understanding of your position and orientation, as well as the features and objects around you. This requires a detailed 3D map of interiors: Homes, businesses, and public spaces. There is currently no available service based dataset of interior 3D features that devices can use to localize against.

While it is technically challenging to build a 3D spatial service, it's clear what needs to happen to build this from a developer standpoint: Platforms need to offer developers a 3D high resolution version of iPhone's Location Services. A fusion of GPS, camera, and inertial measurement unit (IMU) data to determine your exact position and orientation anywhere in the world, inside or outside. You can think of the end result as similar to a massively multiplayer online role playing game, but instead of synchronizing a game world, we're synchronizing 3D points in physical space (features) between users.

What is perhaps more challenging than building the technology for a planet scale spatial service (the AR Cloud) is building the business strategy for

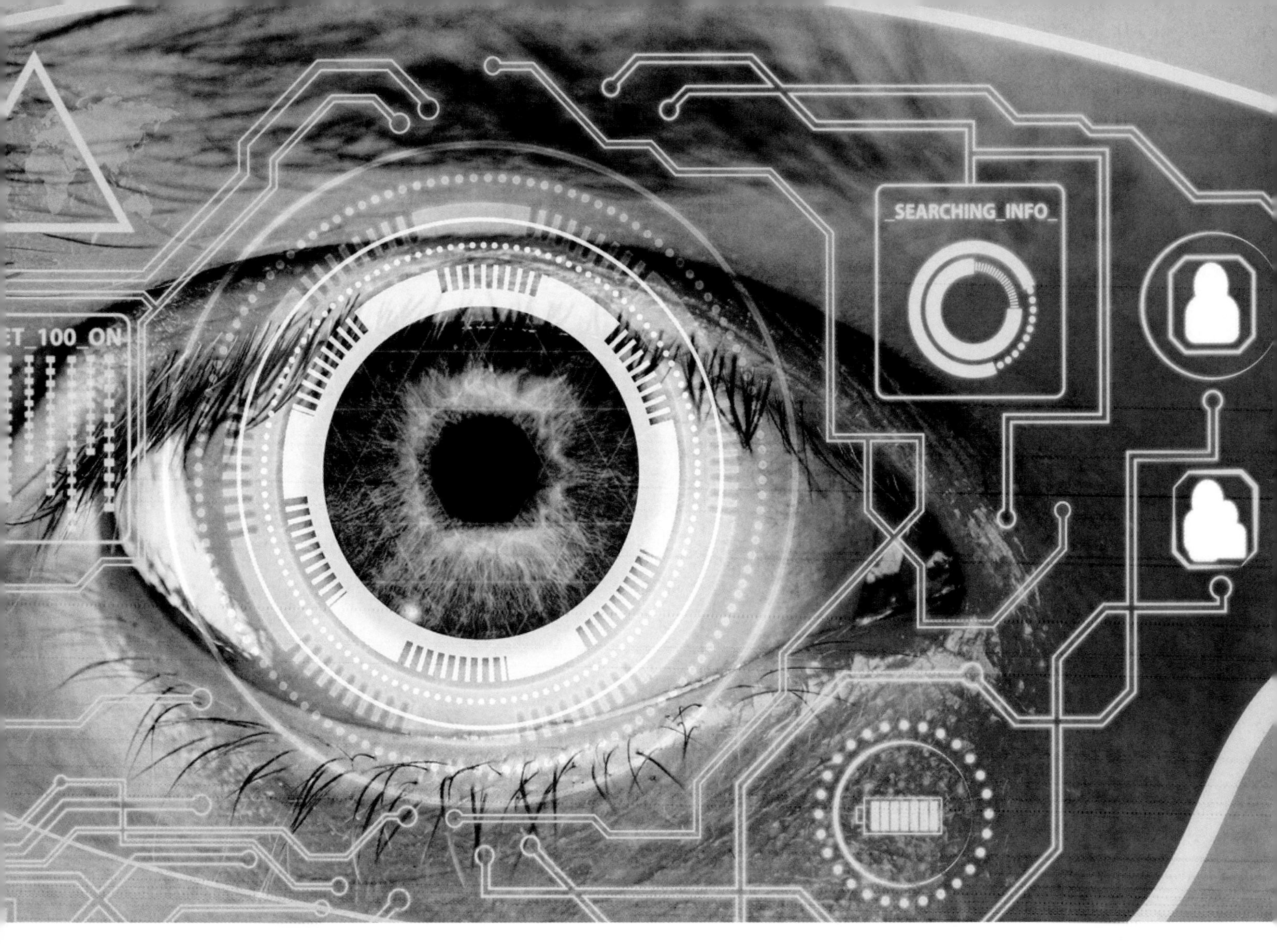

generating and collecting this data set in the first place. Unlike Google Maps, where Google sent employees with mapping equipment into the world to map roads by driving them, a company can't simply send employees into your home or your workplace to map the interior. While this might be possible for public interiors like malls and theme parks, it won't work for privately owned businesses, much less private residences. We have to find compelling reasons to convince consumers and businesses to scan their interiors and store this data on somebody's cloud service.

For consumers, the most likely way will be entertainment applications such as games. As you move through your space playing an AR game, you'll map and refine the features of your home. Currently this is saved on your device, but it needs to be stored on a server to allow for an unlimited amount of mapped spaces, as well as synchronization between different AR devices in the same physical space. For businesses, it seems likely there will be a mix of marketing and advertising use cases geared at consumers in the space, as well as enterprise use cases within the business itself.

This is a great opportunity for startups as this problem is largely unsolved. The big tech companies are only just gearing up to solve this, and most importantly, the value proposition of why people should use AR is just starting to be explored via Apple's ARKit and Google's ARCore. It's only a matter of time until a startup creates a business advertising platform that uses ARKit/ARCore to make it easy for businesses to offer 3D/AR promotions, and in exchange, receive valuable interior mapping data. The next Pokémon GO viral sensation that can be played inside will generate massive amounts of interior residential data that will be incredibly valuable.

Artificial Intelligence

For AR to operate at its full potential, it's not enough for devices and applications to know which place they are in and what general features are in the space. We have to have a more detailed understanding of the individual objects in the space, how they relate to each other, and how they relate to the user. This will be facilitated by artificial intelligence (AI) systems that excel at objection detection and recognition. The latest AI techniques have made it possible for machines to outperform humans in certain visual recognition tasks.

To see projections like this you need more than a very smart wrist band and a projector. You need a projection surface or AR glasses. Scientists are experimenting with wrist projectors that use the skin itself as a screen.

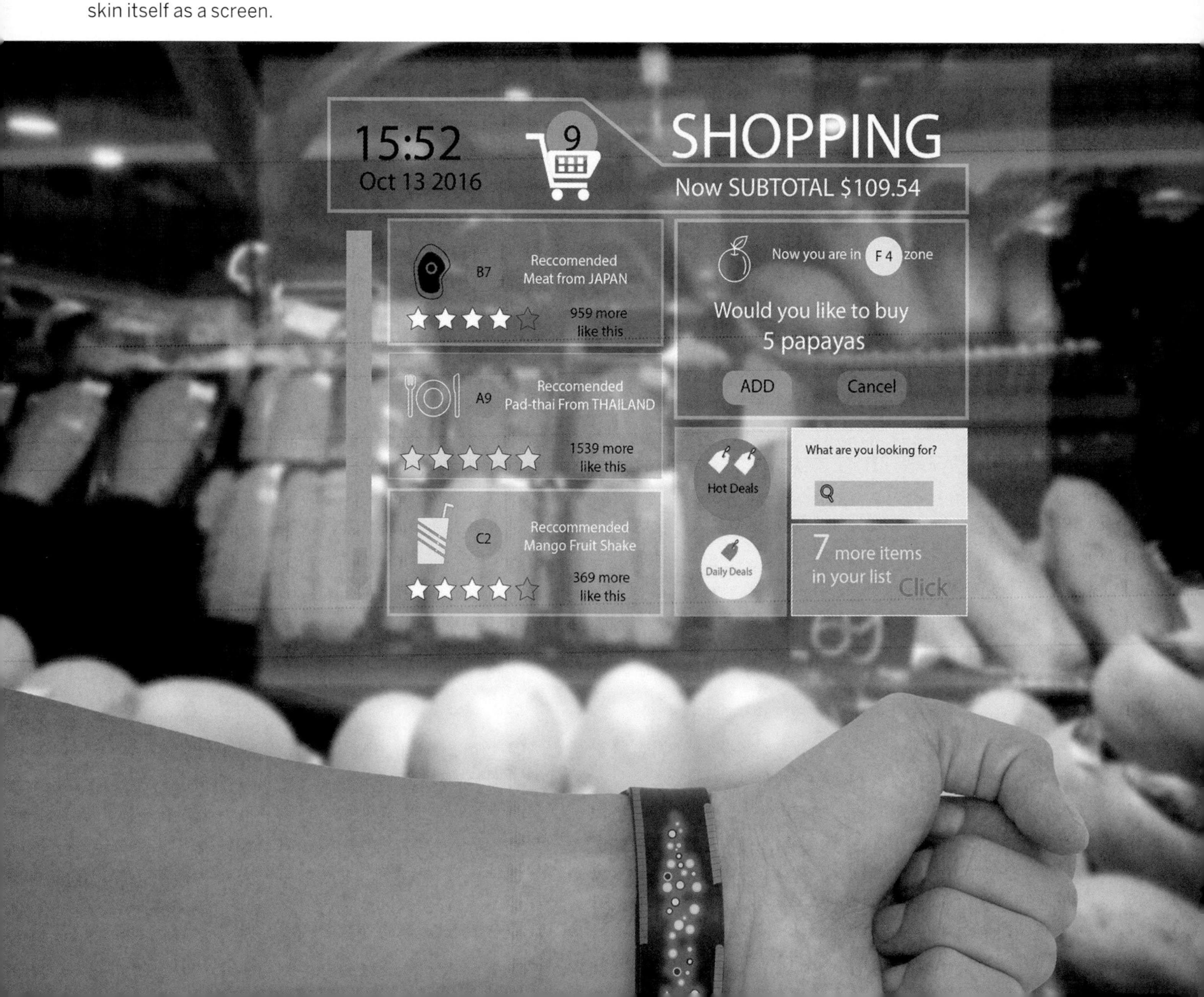

Self driving vehicles use artificial intelligence to identify moving objects, and make predictions about their trajectory and velocity.

If I'm shopping in the grocery store, I may want to run a visual product search while I shop, to let me know if there are better prices available elsewhere in the store, in another store, or online. For this to work, my device has to quickly and reliably understand what the product I am looking at is. Amazon has made tremendous progress in this area.

A more general visual search is a highly ambitious problem that companies including Google, Pinterest, and Blippar are working on. This would allow the device to recognize all types of objects. It could recognize a city bus, and automatically perform a search against bus routes to see which bus this is, and if it's the bus you need to take to the next appointment in your calendar. It could recognize the door to your house, and unlock the door for you as you approach. This is the type of highly context aware AR we've been imagining in books and movies for years now.

This is an opportunity for companies with expertise in machine learning that have access to, or can generate, a large visual training data set to build these types of recognizers. In addition, we will see more general AI that takes the inputs from an AR wearable and combines them with personal data (calendars, preferences) to determine your current context in real time and push relevant data to you as you need it. This sort of on-demand virtual assistant will further add to the seamless and natural interaction with our digital worlds and selves.

User Interface

The ideal AR interface does not require a controller, wand, or other mechanical input device. It should utilize the controllers we all instinctively know how to use: Our hands. Apple pioneered skeuomorphic design, which related 2D user interfaces to physical analogies we were already familiar. AR will be best when we can interact using the 3D analogies of touch and gesture we are already understand. This is a major drawback of ARKit/ARCore. While it offers a glimpse into the AR world, we still interact using yesterday's mobile computing gestures.

A great example of this kind of interaction comes from Leap Motion. Their device allows natural hand tracking, even when your hands are out of view. Furthermore, they've demonstrated some human computer interface paradigms that make a lot of sense for AR.

There is a great opportunity here for companies to build cross platform user interface frameworks (think Qt or Cocoa Touch for AR) that discover and evangelize the user interaction paradigms that will become standard in the future. Some user interfaces will be dependent on the specifics of the hardware platform and its sensing capabilities, but others will be universal across platforms. Finding these universal 3D interaction paradigms, and packaging them in a SDK ready for developers to utilize could be valuable. Using a Leap Motion tracker in VR would be a way to start building the AR interface systems of tomorrow, today.

Hardware

Lastly, we need a form factor or device better than the "magic window" provided by our smartphones. AR hardware has long been envisioned as a wearable of some kind, usually depicted as glasses. We currently have first generation wearable AR displays from Microsoft, DAQRI, and Meta. These are expensive, consumer unfriendly, and in some cases require a tethered computer to run. For the most part, the current generation of AR wearables are geared for enterprise usage.

ARkit and ARCore are attempting to fill the role of a consumer AR wearable by unlocking a subset of AR use cases with today's smartphone. This is a smart approach, as it allows developers previously experienced with 2D development (web and mobile) to get used to 3D development (Unity, Unreal), and allows developers to start thinking about what kind of applications make sense in this paradigm, and what kind of services are needed to enable those applications.

Magic Leap has raised over a billion dollars and has long been rumored to be developing a more practical wearable that could be used by consumers.

The hardware has a very long way to go, and hardware is also notoriously expensive to develop, making this area a challenging target for startups. However, once the hardware, spatial services, AI/object recognizers, and natural user interfaces are combined into one device and SDK, the computing world as we know it will change forever. It's at this point that everyone can start experimenting and building the new use cases and applications that will bring true value to users.

One of the most fundamentally exciting opportunities of any new platform or computing paradigm is finding the use cases that haven't yet been discovered. Beyond entertainment and gaming, the utility of AR is likely to be it's biggest contribution.

Conclusion

Based on the hyped and subsequently slow development of consumer VR, we should be careful to ensure that AR is not a solution in search of a problem. Rather than banking on the novelty of new hardware capabilities ("Wow, I can look around" or "Wow, that animal/person/object is so close to my face!"), we should begin with the end in mind. What is the value proposition that makes AR so useful? What kinds of applications or content makes sense in this new computing paradigm? What tasks can we do only in AR, that just don't make sense in any other format?

Having a platform that is always on and continuously aware of your spatial context and the objects in it will enable powerful AI that understands you well enough to push the data and information you need in real time. Advanced hardware displays will render this information seamlessly and naturally into the real world around you. Finally, you can interact with this information in an intuitive and natural way. This platform will be transformative for businesses and consumers alike, and offers huge opportunities for startups and later stage companies willing to take a risk developing the systems required to make this happen. ■

How Are People Making Money... OR WILL THEY

The Consumer Revenue Models of Virtual Reality

By Stephanie Llamas

There is a debate around VR's status as transformative medium or gimmick. Stakeholders use revenue and audience figures to measure success, though this is clearly premature. There are only 43 million global unique VR users in 2017, so the audience for any specific experience is small. The market will reach just over $2 billion by the end of 2017, dwarfed by the $16 billion music industry, $39 billion in global box office sales, and $98 billion digital games market.

Monetization Type	Definitions	Examples
Upfront/Premium	Users pay once to own an entire piece of content with no additional costs thereafter	Music on the iTunes Marketplace, The Legend of Zelda: Breath of the Wild for the Nintendo Switch
Paymium	Users pay to own a piece of content with the option to purchase add-ons	Games like Grand Theft Auto V and Star Wars: Battlefront 2
Free-to-Play/Freemium	Users get free access to content with the option to purchase add-ons	Games like Candy Crush Sage and League of Legends
Ad-based	Advertisers subsidize content by paying to display ads (the content can be a free or paid service for users)	YouTube, websites
Subscription	Users pay a fee on a regular schedule for unlimited access to content	Netflix, Spotify, Viveport, PC Cafes, cable bundles
Donations	Users have the option, but are not obligated, to donate money for content that is accessible to all users (even if they do not donate)	NPR, Wikipedia
Crowdfunding	Users donate money to a project and can receive access to content or premium services/products for their donation	Kickstarter, GoFundMe, Patreon
Location Based Media (LBE)	Visitors pay a time-based fee or ticket cost to have an out-of-home experience	Movie ticket sales, arcades, theme parks

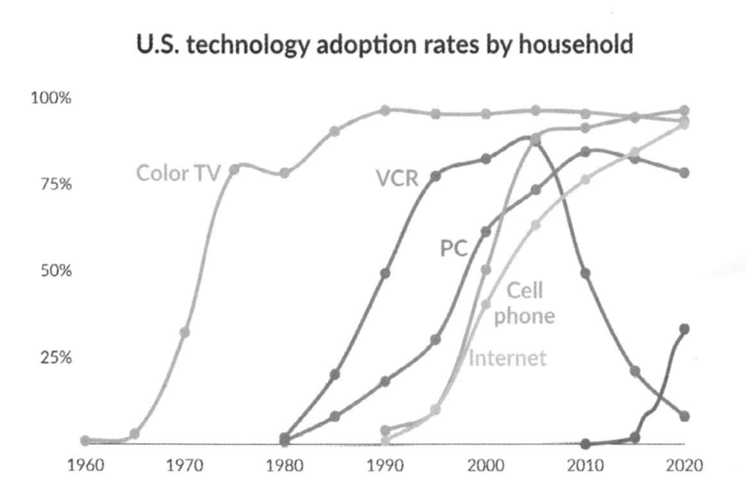

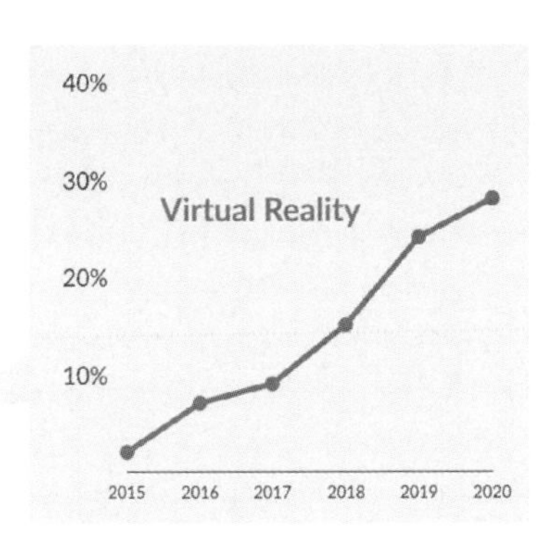

Sources: Michael Felton, The New York Times; Pew Research Center; Gallup; U.S. Census

Adoption rates for different media show us that new communication tools follow the same trends as the ones that came before. Their revenue growth follows a predictable pattern: Slow to startup, and then a quick upswing when they hit an inflection point. If we use revenue as a proxy for where we are in the adoption of this medium, we are still clearly in the preliminary gap, although there is a growing number of indications that this will end soon. Venture activity is up. Relatively inexpensive standalone headsets are now coming to market. And the VR ecosystem continues to scale at a remarkable pace.

So the question isn't if VR can monetize, but when and how. The answer isn't straightforward, particularly because how companies earn revenue is not based on a medium, but instead that medium's content and consumer preferences. Essentially there is no one size fits all method for any platform, and before VR companies can understand how consumers want to spend, they need to cater to how consumers want to consume.

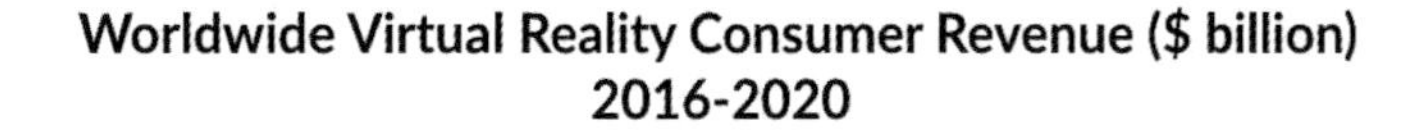

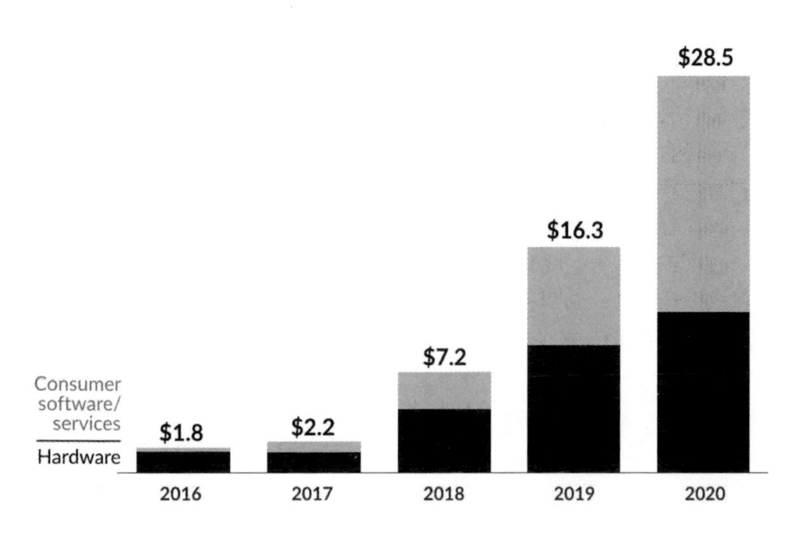

Free-To-Play

The first step to knowing how VR should attack the question of monetization is to look at how its predecessors have done the same. Mobile games are an excellent example of how consumers determine which methods content creators will find success. Not only did it take mobile companies roughly half a decade to find the right model, consumers went from accepting the obligation to spend to demanding free content that gave them the choice. This is how free-to-play disrupted media's traditional revenue models.

Angry Birds is arguably the first smartphone success, catapulting to $200 million in revenue and over 260 million users by 2012, just two years after its launch. During the game's initial growth, consumers were just beginning to understand how to use smartphones but didn't know yet how to consume content, let alone how they wanted to pay for it. In its first year, Angry Birds made $1 million from advertising in their free version of the app, but made six times that from paid downloads.

It makes sense that consumers would accept this type of monetizing since consumers have been conditioned by the console and PC gaming industries to pay for a full game up front. The frictionless e-commerce environment greatly lubricated this adoption. All users had to say is "yes" and the App Store seamlessly charged their credit card.

However, upfront spending eventually presented a challenge both to consumers and game developer Rovio since consumers weren't getting the chance to try content before committing dollars to it (so what if it's not worth the cost?). And Rovio had a revenue ceiling, only able to monetize a player once per release. Rovio's games revenue steadily declined over the following years once free-to-play began gaining steam. In 2014, Rovio's entire catalogue of games made roughly $90M. Meanwhile, free-to-play standout Candy Crush Saga made $1 billion that year on in-game spending.

Although free-to-play publishers don't earn revenue from every player, play-

Before VR companies can understand how consumers want to spend, they need to cater to how consumers want to consume.

ers who want to customize their experience by, for example, moving through levels quicker, or eliminating ads, can pay to do it. In 2014, Clash of Clans, which remains a top 10 grossing app, only monetized 4.4% of players. But each paid an average of $25.35 a month, far more than consumers are willing to spend upfront for an app. It was the second highest grossing mobile game that year, earning $1.3 billion (more than half the entire VR market in 2017). It still makes roughly $100 million a month, with an accessible audience of roughly 2 billion worldwide smartphone users.

As of November 2017, all of the top 10 grossing mobile games in both the App Store and Google Play are free-to-play games. Free-to-play even dominates on PC with MOBA League of Legends continuing its long standing reign as the highest grossing game worldwide, earning over $2 billion in 2017 alone. The game doesn't even promise players they can pay to win. In-game currency that players buy with real money (Riot Points) can only be used to purchase items that do not directly impact performance (such as cosmetic items), and only currency earned by playing matches (Influence Points) can be used to buy things like power boosts.

But can free-to-play work for VR? The short answer is, not yet. Publisher CCP is one of the first, and only, companies to use in-game transactions in their VR games as of 2017. However, this wasn't enough to sustain their popular games on the platform, causing them to shut down two of their studios and refocus development on non-VR PC and mobile games. Because only a small percentage of players spend money on any given free-to-play title, the user base needs to be sizeable. League of Legends doesn't just make more money than any other PC game, it has the most players with 84 million average monthly users.

Not only does free-to-play require a user base that VR doesn't have yet, it is only as effective as its payments process. Part of the reason players spend in games like League of Legends is because it takes less than a minute to acquire Riot Points. Likewise, the only way VR content can earn free-to-play revenue is with frictionless payments. Companies like Payscout and Worldpay have created solutions that promise a seamless checkout experience with support from Mastercard and Visa. Now that major providers are thinking of ways to facilitate payments in VR, the only thing that's missing is the volume of users needed to make free-to-play work, and experiences with enough content and replayability to keep users coming back for more.

Advertising

While free-to-play monetization is a fairly new concept, many media have taken lessons from their predecessors, focusing primarily on ad-subsidized content, subscriptions or upfront purchases. These models are pretty straightforward. Consumers can either 1) agree to tolerate ads in exchange for subsidized or free content, or they can 2) pay to avoid advertising altogether, possibly getting access to premium content as well. YouTube, for instance, relies on advertising to subsidize their service, while music streaming services Spotify and Pandora offer both ad-subsidized and subscription services. Meanwhile iTunes provides a marketplace with paid content that consumers purchase upfront, download and keep forever.

These tried and true models are becoming less relevant as consumers pay for customizable services. For instance, cord cutters are tired of paying for channels with predetermined programming schedules, especially sports channels, which over inflate the cost of a subscription, even if viewers don't want them. Consumers have agency when it comes to how they spend on media, so tolerating ads is no longer a given. That means companies must accommodate to the monetization preferences of the consumer, not the other way around.

Of course, free doesn't always mean free. But you can pay more for a better experience. Sometimes free still just means free, but it's mostly thanks to advertising. YouTube provides more than 1 billion users a completely free service that exclusively earns revenue from advertising.

But more services are giving consumers the option to pay their way out of ads, which is why YouTube has created an ad-free subscription service called YouTube Red. Other streaming services like Netflix offer no (or minimal) advertising for a premium subscription, which earned the service $8.8 billion in 2016 as a result of this kind of demand.

Dreamscape may be able to charge as much as $3 a minute at peak **for an indescribable free roam VR experience** made by some of **Hollywood's greatest talents**

This isn't necessarily a global trend. Chinese mobile users are more permissive of ads if it means completely free content. Whereas in the United States, 7% of mobile players purchased in-game content in 2017, the same goes for just 4% in China.

Before VR can entertain the ad-based model, it needs ads to support content. Most brands that are using VR are doing it as a way to show they can create innovative content, oftentimes publishing their own full experiences. IKEA, Gatorade and the Hawai'i Tourism Authority are among the brands that have created independent pieces of content, some of which they've actually monetized. This, however, does not fall in line with the ad-based model since brands aren't paying to advertise in other content. That's because advertisers largely rely on impressions, and, again, the VR audience isn't there yet.

There do exist VR ad networks like Immersv, and even analytics platforms like Retinad. These tools will make it easy for brands to get involved once they see potential ROI. It's inevitable since advertising has permeated every entertainment medium we currently use. Once VR adoption in the U.S. hits a third of households in 2019, there will be enough content, impressions, and engagement for brands to justify venturing into this new territory.

Subscriptions

Subscriptions in general are joining free-to-play as a way to drive continual earnings versus the limited revenue that comes from upfront purchases. Essentially, it's the difference between allowing consumers to "borrow" versus own the content. With the introduction of streaming media, subscriptions are increasingly becoming a compromise between providers and consumers, giving users access to a library of content on demand in exchange for continued revenue.

Subscriptions and free-to-play are the main drivers for the development of content-as-a-service (as opposed to the traditional model where content has been seen as a product). This requires providers to serve the needs of their consumers by continuously updating content in order to keep them paying (e.g., game levels, video libraries, application functionality, etc.). Since products are paid for upfront, providers do not need to give continual updates because they have already made as much as they can off of each consumer.

Currently, there is not enough content to fully support the types of distribution channels that can benefit from subscriptions. For instance, although much of the content on an application like Hulu doesn't appeal to everyone, subscribers can still find enough to justify a monthly cost. However, VR distributors like Jaunt and Littlstar have limited catalogues, making it hard for users to find content that keeps them satisfied each month. Once they do, subscriptions can still present a challenge to the creators themselves since creatives are not paid on a per user basis. Currently, subscription payments do not cover the production costs for most VR content, so this model has provided limited opportunities for industry stakeholders.

Subscriptions have more immediate potential for games than video content. A major complaint for VR players is that games can be too expensive for the limited amount of gameplay they can provide. But production costs are so high, those prices are the only chance creators have to recoup their costs. Viveport, for instance, allows gamers to subscribe to the platform and receive access to a rotating catalogue of games for one flat monthly fee. As players have more titles to try, this will be a way for them to do so economically, no matter how long or short the game's content.

LBVR

Added fees to tickets like experiences in museums are one of the few monetization models that are both accepted by consumers and apply well to VR. Location based VR (LBVR) currently has a more straightforward path to VR monetization than at home content, which is why it is the second highest grossing VR software segment this year. What threatens its success, however, is its ability to scale.

Virtual World Entertainment (VWE) networked custom pods based on flight simulator technology and offered them in malls around the world, beginning with the Battletech Center in Chicago in 1991. *Battletech* started as a role playing game, and the richness of the mythology informed the simulations, producing a passionate following who, like this Houston crew, have kept VWE pods working to this day. Photo courtesy Mechcorps

A theme park is one of the few environments made for VR. The tradeoff for high priced entry and hours of standing in line is a world of rides that taps into multiple senses to boost adrenaline and dopamine for a short, but meaningful period of time. This is amplified by tying popular intellectual property into the rides and theme park, transforming a visitor's entire environment into something of an alternate reality (remind you of something?). Theme parks were able to pioneer 3D and 4D experiences because they have the space, money, and demand, which then indirectly boosts revenue. So VR can naturally integrate the uniqueness of its experiences into the uniqueness of a theme park's environment.

However, VR-only experiences that rely on direct per minute revenue, such as arcades and VR-on-wheels (traveling vehicles with VR experiences), face many more challenges. The first is that the current generation of headsets often require time and assistance to put on. This coupled with the fact that

WHY WE NEED

By Ori Inbar

Ori Inbar is the founder of the first early stage fund, Super Ventures, which is dedicated to augmented reality (AR). Ori is a recognized speaker in the AR industry, lecturer at NYU, as well as a sought after adviser and board member for augmented reality startups. In 2010, he co-founded Augmented World Expo, one of the most important conferences in the industry.

With the release of Apple's ARKit and Google's ARCore, creating augmented reality (AR) apps became a commodity overnight. It's free, it's cool, and it just works. So far, the "mass adoption" is mostly among AR developers, and its generating tons of YouTube views. But developers are yet to prove their apps can break through the first batch of novelty apps and gain mass user adoption.

For sure, there will be a couple of mega-hits (e.g. Pokémon Go) that will enjoy the newly made mega distribution channel for AR apps on iPhones and on high end Android phones. But, I do not anticipate hundreds of millions using AR apps all day, every day. ARKit and ARCore based AR apps are like surfing the web with no friends. It's so 1996.

The ARKit, released in September, 2017, is the best thing that happened to the AR industry. In just a couple of days, Apple's AR enabling new iOS 11 was on hundreds of millions of phones. But massive adoption of AR apps will take much more than that. It will happen with the AR Cloud, when AR experiences persist in the real world across space, time, and devices.

I was hoping the iPhone X will introduce a key ingredient for massive AR adoption, not the ARKit. But rather, a back facing depth camera that will put in the hands of tens of millions a camera that senses the shapes of your surroundings, and can create a rich, and accurate 3D map of the world to be shared in the AR Cloud. That hasn't happened yet. Hopefully it will in the next version. This will only slow down the creation of the AR Cloud, but it cannot stop the demand for it.

A real time 3D (or spatial) map of the world, the AR Cloud, will be the single most important software infrastructure in computing. Far more valuable than Facebook's social graph or Google's PageRank index. In a nutshell, with the AR Cloud, the entire world becomes a shared spatial screen, enabling multi-user engagement and collaboration.

The AR Cloud is a shared memory of the physical world and will enable users to have shared experiences, not just shared videos or messages. It will allow people to collaborate in play, design, study, or team up to problem solve

AN AR CLOUD

Courtesy Microsoft

anything in the real world. Multi-user engagement is a big part of the AR Cloud. But an even bigger promise lies in the persistence of information in the real world.

We are on the verge of a fundamental shift in the way information is organized. Today, most of the world's information is organized in digital documents, videos, and information snippets stored on a server, and ubiquitously accessible on the net. But it requires some form of search or discovery.

Based on recent Google stats, over 50% of searches are done on the move (searched locally). There is a growing need to find information right there where you need it, in the now. The AR Cloud will serve as a soft 3D copy of the world and allow you to reorganize information at its origin, in the physical world (or as scientists call it in Latin: in situ). With the AR Cloud, the how to use of every object, the history of any place, the background of any person, will be found right there, on the thing itself. And whoever controls that AR Cloud could control how the world's information is organized and accessed.

When Will This Happen?

The AR Cloud is not for the faint of heart. Industry leaders are imploring developers, professionals and consumers alike to have patience as it'll take a while for this to materialize. But as investors in AR and frontier technology, this is the time to identify the potential winners of this long distance race.

In the past decade, several companies have been providing AR services from the cloud, starting with Wikitude as early as 2008. Then Layar, Metaio (Junaio), and later: Vuforia, Blippar, Catchoom, among new entrants. But these cloud services are typically one of two types: (1) Storing GPS or location related information (for displaying a message bubble in a restaurant) or (2) Providing image recognition services in the cloud to trigger AR experiences.

These cloud services have no understanding of the actual scene and the

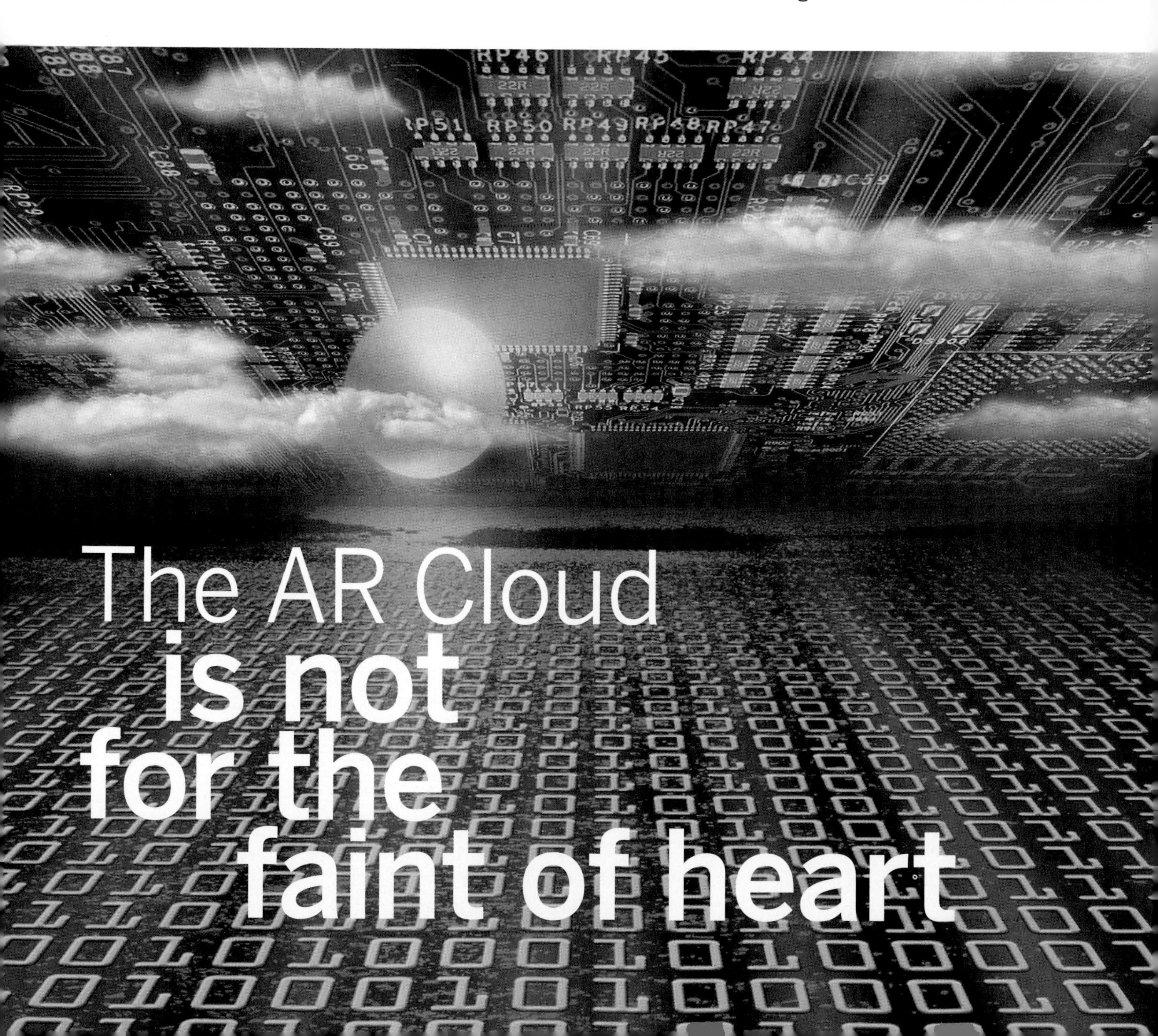

The AR Cloud can be thought of as a machine readable **1:1 scale model of the real world.**

geometry of the physical environment. And without it, it's tricky to blend virtual content with the real world in a believable way, let alone share the actual experience (not just a cool video) with others.

Imagine Pokémon Go With The Cloud

The incredible reach of Niantic's Pokémon Go was a fluke. An outlier. A one of a kind. It will be super hard, if not impossible, to replicate its success with similar game mechanics. Though, we are about to find out: Niantic is preparing to launch a similar massive AR game based on the Harry Potter mega franchise in the first half of 2018. For Pokémon Go, game servers store geolocation information, hyperlocal imagery, and players' activity. but not a shared memory of the physical places in which its 65 million monthly active users are playing. Thus, no real shared experience can occur. For that it would need the AR Cloud.

Several scientists have conceived different aspects of the AR Cloud since the 1990's, but I have yet to see a concise description for the rest of us. So here is my simplified version in the context of the AR Industry.

An AR Cloud System Should Include: A Persistent Point Cloud

A point cloud as defined in wikipedia is "a set of data points in some coordinate system (x,y,z)," and by now is a pretty common technique for 3D mapping, reconstruction, surveying, inspection, and other industrial and military uses. Capturing a point cloud from the physical world is a "solved problem" in engineers' jargon. Dozens of hardware and software solutions have been around for a while for creating and processing point clouds using active laser scanners such as LiDAR. Depth or stereo cameras such as Kinect, and monocular camera photos, drone footage or satellite imagery processed with photogrammetry algorithms. And even using synthetic aperture radar systems (radio waves) such as Vayyar or space borne radar.

For perspective, photogrammetry was invented right along with photography. So, the first point cloud was conceived in the 19th century. To ensure it has the widest coverage and always keeps a fresh copy of the world, a persistent point cloud requires another level of complexity: The cloud database needs to have a mechanism to capture and store a unified set of point clouds fed from various sources (including mobile devices) and its data needs to be accessible for many users in real time.

The solution may use the best native scanning or tracking mechanism on a given device (ARKit, ARCore, Tango, Zed, Occipital, etc.), but must also store the point cloud data in a cross platform accessible database.

The motivation for users to share their personal point clouds will be similar to the motivation Waze users have: Receive a valuable service (optimal navigation directions), while in the background sharing information collected on your device (your speed at any given road segment) to improve the service for other users (update timing for other drivers).

Nevertheless, this could pose serious security and privacy concerns for AR mapping services, since the environments mapped would be much more intimate and private than your average public road or highway. I see a great opportunity for crypto-point-cloud startups.

Google Tango calls it "Area Learning. Which gives the device the ability to see and remember the key visual features of a physical space (the edges, corners, other unique feature), so it can recognize that area again later." The AR Cloud's requirements may vary indoors and outdoors, or whether it's used for enterprise or consumer use cases. But the most fundamental need is for fast localization.

Instant Localization From Anywhere

Localizing means estimating the camera pose which is the foundation for any AR experience. In order for an AR app to position virtual content in a real world scene, the way it was intended by its creator, the device needs to understand the position of its camera's point of view relative to the geometry of the environment itself.

To find its position in the world, the device needs to compare these key feature points in the camera view with points in the AR Cloud and find a match. To achieve instant localization, an AR Cloud localizer must narrow down the search based on the direction of the device and its GPS location (as well as leveraging triangulation from other signals such as wifi, cellular, etc.). The search could be further optimized by using more context data and AI.

A REAL TIME 3D (OR SPATIAL) MAP OF THE WORLD THE AR CLOUD, WILL BE THE SINGLE MOST IMPORTANT SOFTWARE INFRASTRUCTURE IN COMPUTING

Once a device is localized, ARKit or ARCore, using the inertial measurement unit (IMU) and computer vision, can take over and perform the stable tracking. As mentioned before, that problem is already solved and commoditized.

Dozens of solutions exist today to localize a device such as Google Tango, Occipital Sensor, and of course ARKit and ARCore. But these out of the box solutions can only localize against a local point cloud, one at a time. Microsoft HoloLens can localize against a set of point clouds created on said device. But, it can't localize against point clouds created by other devices.

The search is on for the "ultimate localizer" that can localize against a vast set of local point clouds from any given angle and can share the point cloud with multiple cross platform devices. But any localizer will only be as good as its persistent point cloud, which is a great motivation to try and build both.

Place and Visualize Virtual Content in 3D

Once you have a persistent point cloud and the ultimate localizer, the next requirement to complete an AR Cloud system is the ability to position and visualize virtual content registered in 3D. "Registered in 3D" is the technical jargon for "aligned in the real world as if it's really there". The virtual content needs to be interactive so that multiple users holding different devices can observe the same slice of the real world from various angles and interact with the same content in realtime.

The AR Cloud landscape spells the biggest opportunity in AR to grow a Google-size startup since Magic Leap. Speaking of Magic Leap. It used to have the ambition to build the AR Cloud. Do they still have it? Should we care? ■

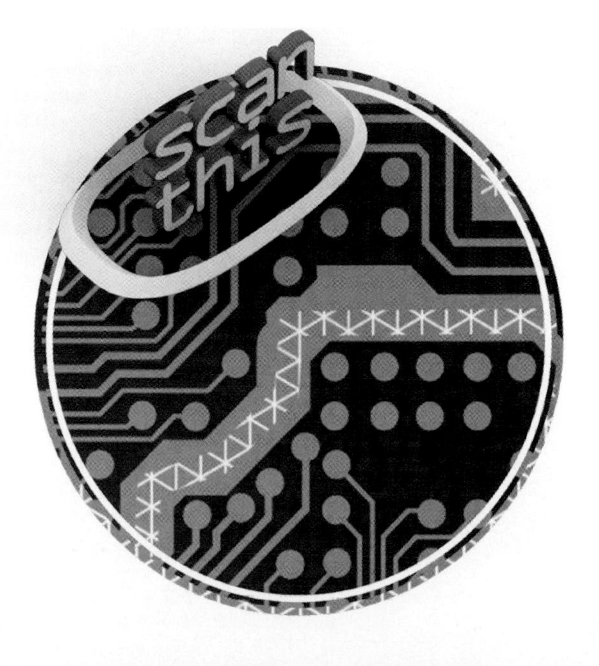
scan
this

Why The AR CLOUD Will Be Bigger Than Search

by Matt Miesnieks

Matt Miesnieks is a serial AR entrepreneur and investor in early stage emerging technology startups. He is CEO of 6D.ai, which is working on the AR Cloud, and a partner at SuperVentures.

If you were asked what is the single most valuable asset in the technology industry today, you'd probably answer that it's Google's search index, or Facebook's social graph or maybe Amazon's supply chain system. I believe in fifteen years time, there'll be another asset at least as valuable as these: The AR Cloud. That doesn't exist today.

Will one company eventually own the AR Cloud? History says probably. Will it be a new company? Also probably. Just as it was hard to imagine Microsoft losing its position in 1997, it's hard to imagine in 2017 Google or Facebook losing their position. But nothing is guaranteed. I'll try to lay out the arguments supporting each of the three sides playing here (incumbents, startups, open web).

Apple's ARKit was released at the end of September 2017. And Google fired right back with ARCore, replacing Tango, which had been with us since 2016. After much hype and excitement, we have found that despite the surface detection iPhones can now do, and the ease with which ARKit allows us to place a digital object in the world, we've basically already hit the limits of what ARKit can do on iPhones, after only six months. And it's not much. The key piece of infrastructure needed to make the phone detect places and people, context, is missing.

What Is The AR Cloud?

To get beyond ARKit and ARCore, we need to start thinking bigger than ourselves. How do other people on other types of AR devices join us and communicate with us in AR? How do our apps work in areas bigger than our living room? How do our apps understand and interact with the world? How can we leave content for other people to find and use? To deliver these capabilities, we need cloud based software infrastructure for AR. In the previous part of this chapter, my Super Ventures partner Ori Anbar, co-founder of Augmented World Expo (AWE), refers to all this stuff as the AR Cloud.

The AR Cloud can be thought of as a machine readable 1:1 scale model of the real world. Our AR devices are the real time interface to this parallel virtual world, which is perfectly overlaid onto the physical world.

It's hard to imagine in 2017 Google or Facebook losing their position. But nothing is guaranteed.

Why all the "meh" from the press for ARKit and ARCore ?

When ARKit was announced at Apple's World Wide Developer Conference (WWDC) this year, CEO Tim Cook touted augmented reality, telling analysts: "This is one of those huge things that we'll look back at and marvel on the start of it." A few months went by. Developers worked hard on the next big thing. But, the reaction to ARKit at the iPhone launch keynote was "meh". Why was that?

It's because ARKit and ARCore are currently at version 1.0. They only give developers three very simple AR tools: (1) The phone's 6DOF pose, with new coordinates each session; (2) a partial and small ground plane; and (3) a simple average of the scene lighting.

In our excitement over seeing one of the hardest technology problems solved, a robust 6DOF pose from a solid visual inertial odometry (VIO) system, and Tim Cook saying the words "augmented" and "reality" together on stage, we overlooked that you really can't build anything too impressive with just those three tools. Their biggest problem is people expecting amazing apps before the full set of tools to build them existed. Note that it's not the if, but the when, that we've gotten wrong.

How To Make AR Great

Clay Bavor referred to the missing pieces of the AR ecosystem as connective tissue. An AR app needs some interaction or connection with the real world. With physical people, places or things. Without this connection, it can never really be AR-Native. These capabilities are only possible with the support of the AR Cloud.

How do we support multiple users sharing an experience? How do we see the same virtual stuff at the same time, no matter what device we hold or wear, when we're in the same place or not? You can choose a familiar term to describe this capability based on what you already know, i.e. "multiplayer" apps for gamers, or "social" apps or "communicating" apps. It's all the same infrastructure under the hood and built on the same enabling technology: Really robust localization, streaming of the 6DOF pose and system state, 3D mesh stitching and crowdsourced mesh updating, are all technical problems to be solved. Also, don't forget the application level challenges like access rights, authentication, etc.

How do AR apps connect to the world and know where they really are?

CPS just isn't a good enough solution, even with the forthcoming GPS that's accurate to 1 foot. How do we get AR to work outside in large areas? How do we determine our location both in absolute coordinates (Lat/Long) and relation to existing structures to sub-pixel precision? How do we do achieve this both indoors and out? How do we ensure content stays where it's put, for days or years later? How do we manage so much data? Localizing against absolute coordinates is the really hard technical problem to solve here.

How do our apps understand both the 3D structure or geometry of the world (the shape of things)? I.e. That's a big cube like structure my Pokémon can hide behind or bounce into, and identify what those things actually are? I.e. The blob is actually a couch and my virtual cat should stay off couches. The challenges include:

- Real time on-device dense 3D reconstruction
- Real time 3D scene segmentation
- 3D object classification
- Backfilling local processing with cloud trained models

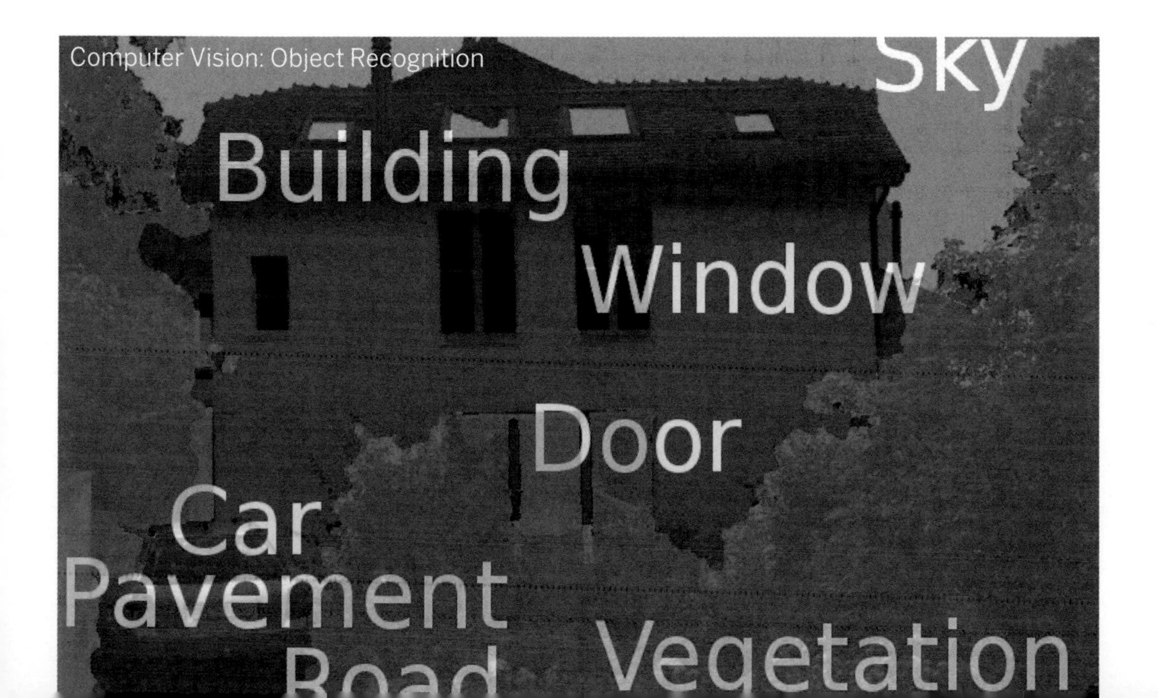

Like much in AR, it's not that hard to build something that demonstrates well, but it's very hard to build something that works well in real world conditions.

Is today's cloud up to the job?

When I worked in telecom infrastructure, there was a little zen-like truism that said "there is no cloud, it's just someone else's computer". We always ended up working with the copper pairs, fibre strands or radio spectrum that physically connected one computer to another, even across the world. It's not magic, just difficult. What makes AR Cloud infrastructure different from the cloud today (powering our web and mobile apps) is that AR (like self driving cars, drones and robots) is a real time system. Anyone who has worked in telecom deeply understands that real time infrastructure and asynchronous infrastructure are two entirely different beasts.

So while many parts of the AR Cloud will involve hosting big data, serving web APIs, and training machine learning models (just like today's cloud), there will need to be a very big rethink of how do we support real time applications and AR interactions at massive scale. Basic AR use cases like:

- Streaming live 3D models of our room while we "AR Skype".
- Updating the data and applications connected to things, and presented as I go by on public transport.
- Streaming rich graphical data to me that changes depending on where my eyes are looking, or who walks near to me.
- Maintaining and updating the real time application state of every person and application in a large crowd at a concert.

Without this type of UX, there's no real point to AR. Otherwise, let's just stick with smartphone apps.

But, supporting this for eventually billions of people is a huge opportunity. 5G networks will play a big part and are designed for just these use cases. If history is any guide, some, if not most of today's incumbents who have massive investments in the cloud infrastructure of today, will not cannibalize those investments to adapt to this new world.

Is ARKit (or ARCore) useless without the AR Cloud?

Ultimately, it's up to the users of AR apps to decide this. Useless was a provocative word choice. As of the end of Q4 2017, based on early metrics, users are leaning toward "almost useless". My personal belief is that useful apps can be built on today's ARKit and ARCore, but they will only be useful to some people, occasionally. They might be a fun novelty that makes you smile when you share it. Maybe if you're buying a couch, you'll try it in advance. But these aren't the essential daily use apps that define a new platform. For that we need AR-Native apps. Apps that are truly connected to the real world. And to connect our AR Apps to each other and the world, we need the infrastructure in place to do that. We need the AR Cloud. ■

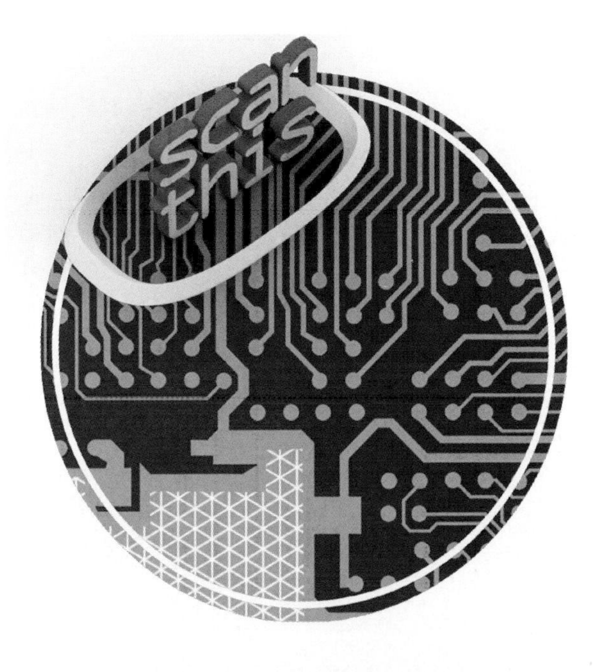

The YouTube of AR Won't Be YouTube

by Matt Miesnieks

A couple of years ago when I was working at Samsung, around the time the GearVR was being launched, there were lots of ideas flying around regarding the services to offer alongside the device.

Many (most) of these ideas were related to hosting 360 video content, and our boss David Eun (ex-YouTube), often reminded everyone that YouTube will be the YouTube of VR. He meant that the content on this new VR platform gave a similar enough experience to existing video content, that a new service couldn't compete with the incumbent. It turned out he was pretty much right.

I don't think this is going to be the case for AR. I think there's a genuine difference in that AR represents a new mass medium, not a new form of a current medium.

This means that the iTunes of AR, quite possibly won't be iTunes. It will be an AR-Native application that changes the way we experience art and entertainment. iTunes may adapt, but that's not guaranteed at all.

So why is AR different to VR Video?

The difference is that VR video is essentially the same old 2D video we know and love, but super duper wide screen. So wide we have to look sideways to see it all. It isn't a new thing. AR is a new thing. For the first time, the media is experienced as part of the real-world. Context is the new attribute.

It's multi-sensory, dynamic, interactive and now can be in my living room, or the street. And a change in context changes the experience entirely. To illustrate the difference: If you experience Star Wars in the cinema, or even VR, you are "escaping" to another galaxy where you are immersed in that universe. But it's been a real challenge to bring Star Wars to AR in a way that's not a novelty. Because you need to somehow deal with the cognitive dissonance of "why is R2D2 in my kitchen"?

The context changes the experience.

If we look at music, the way we experience music will potentially change due to AR, in a similar way that music changed when recordings were invented, or the walkman. In one stroke, the context in which we experienced music changed entirely. Inventions like the iPod, radio or streaming didn't really change the medium beyond adding quantity. But, going from live to take home (gramaphone), or from at home to out and about (walkman), changed the experience entirely, because the context in which we experienced the music changed.

With AR, we're going to experience music and art (static or interactive) in context with our lives. AR can adapt the real world to let us inhabit the emotional landscapes we imagine when we hear our favorite track. Katy Perry fans may see their world literally become a little more neon when her music comes on. The bass line from a song that happened at a significant time of our lives may subtly play in the background when we are near that place, or if similar contextual conditions are triggered (a date and time, or a person nearby).

I think there's a genuine difference in that **AR represents a new mass medium,** not a new form of a current medium.

AR and Blockchain have the potential to re-enable scarcity in the digital world. This is one of the most powerful economic disruptors that AR will enable, and still very few people are thinking deeply about it. This could play out through certain types of digital street art starting out pixelized and only becoming "hi-res" after a certain number of people see it. Or the reverse, and it can decay over time as more people see it. Only the first fifty get the full experience. Digital paintings (or sculptures) can be cryptographically certified as a limited edition "original" and all copies are degraded slightly.

When we think about music or art and context, there's an example that we've all experienced. Compare the difference between listening to music at home versus sitting on a beach overlooking the sunset and choosing a track that's perfect for that moment. That's the way in which context is a part of the experience, and emotionally improves it. And in a small way, the resulting experience is a collaboration between you and the artist.

Katy Perry fans may see their world literally become a little more neon when her music comes on.

AR takes this to a whole new level. An AR device will have a greater awareness of the real world than any smartphone can have. This means that the ability to match (either automatically or manually) a song or image or visual effect to the moment, is far greater than just selecting a track from a playlist.

Artists are now able to give more control over to the audience, so the experience can be far more personal. This could involve the stems, connecting sounds or visuals from an album to individual objects, or subtly changing the ambience of an entire room. Imagine an iPhoneX style face mask, but applied to your walls and ceiling. We have added context in the past through making a mix tape (async creation) or a Spotify playlist/feed (near real time), but when the context becomes applied in real time and is shared, which is native to AR, then new forms of collaborative expression emerge.

I believe it opens the door for a new type of open source movement to emerge. For many years a coder (or company) would write the code, and release it under their name. You could buy it and use it, but that was it. Open source meant that creators could publish both entire products as well as the components that made up those products. The general public license (GPL) meant credit was recognized (though payment was problematic pre-bitcoin). Further, when great creators (both famous and anonymous coders) were able to create together, building on each others work, far better products were invented. Today open source software underpins almost the entire Internet.

Today, art and content are created in a very similar way to the closed source model. An artist or label/studio produces the product and we consume it the way we are told. There's very little scope for the audience to apply their own context, or to reuse components into a reimagined expression. Sampling being an exception, and again credit and payments are a challenge to say the least. User generated content (UGC) web platforms went part way toward making creation a two way collaboration. And AR can complete that journey.

From art being a static "created and done" process only involving the artist(s), it now becomes a living process involving artist and audience. The process itself, the system or the code, and the assets, becomes the art.

The one area today where the artist and audience almost create together is at a live concert. This is where the context of the audience, current social environment, weather, etc. all come together to make something unique and greater than the sum of its parts. The massive opportunity for AR is that today there is no way to connect with the audience digitally in that moment of a shared heightened emotional experience.

The crowd all use their generic camera apps, but that information is "single user". The potential to use an AR-enabled smart camera app, that can provide additional layers of context (live data feeds, responsiveness to what others in the crowd or remotely are doing, augmented stage shows, etc.) and then capture all the data around that, and package it into a take home experience is extremely compelling.

The fact that real world context is the defining characteristic of AR, when applied to art and content, it's natural that the difference between the creator and the audience will be far less clear in the future. Like open source software, enabling creators to easily build on each other's work, will mean even greater things are created.

Former Disney animator Glen Keane, who brought The Little Mermaid to life, used Google Tilt Brush to make a dimensional sketch of the adventurous princess.

Who is going to figure this out?

I can tell you who won't figure it out, and that's the marketing arms of labels and studios. Nearly everything i've seen so far regarding music/art in AR has been using the medium as a novelty based marketing channel to promote "the real product" which is a YouTube view or Spotify listen. A Taylor Swift Snap filter, or volumetric video of a piano performance replayed in my bedroom is just trying to squeeze the old medium into the new, like encyclopedias publishing on the web before wikipedia came along. It's dinosaurs facing the ice age.

I think we're going to see some rapid evolution as artists experiment directly in AR. The symbiosis between the artist and the toolmaker is going to be an incredibly important relationship over the next few years. I'm starting to see immersive tool makers replacing steps of existing creative workflows, or artists struggling to express their AR ideas with crude tools, or (sadly) some high profile artists let their labels experiment for them. Occasionally, I'll meet someone like Molmol Kuo and Zach Lieberman from YesYesNo who can both create the tools and the art, and they are doing ground breaking work.

Startups like Cameraiq.co are helping artists connect with new camera centric tools at both live events and in regular life. The Glitch Mob in LA are also pushing new ideas around music and AR, and fusing mediums. Tilt Brush is finding

cracks to escape from VR into AR. My wife, Silka Miesnieks, is also exploring this world of enabling creative and artistic expression through her work leading the Adobe Design Lab. It's a cambrian explosion of ideas right now. When a new medium emerges, like the Web did, it's the artists who define the early interactions and successes, not the MBAs. Fantom is a startup co-founded by Robert (Rob) Del Naja, co-founder of UK's popular trip-hop musical group, Massive Attack, which is reknowned for its ambitious use of multi-media. Fantom sees coders are just as much a part of the artistic process as the musicians or painters. 2018 is going to be an exciting year. I'm excited as an entrepreneur at the prospect of dominant large companies like YouTube or Spotify becoming exposed to new forms of competition for people's hearts and minds.

This means that we will experience the work of the artists we love in newer, deeper, more personal ways, which will enhance our daily lives. The act of creating and experiencing the creations of others is something no AI program will ever replicate.

Thanks to Silka Miesnieks for help preparing this chapter.

UK trip hop group "Massive Attack" is reknowned for its ambitious, large scale multimedia events

The History and Future OF REMOTE COLLABORATION

by Mark Billinghurst

Mark Billinghurst is an augmented reality pioneer who has been conducting research in AR for over 20 years, developing innovations such as the first open source AR tracking library (ARToolKit), the first mobile AR advertising experiences, and the first collaborative AR applications. He is currently a professor at the University of South Australia, and the former director of the world famous HIT Lab.

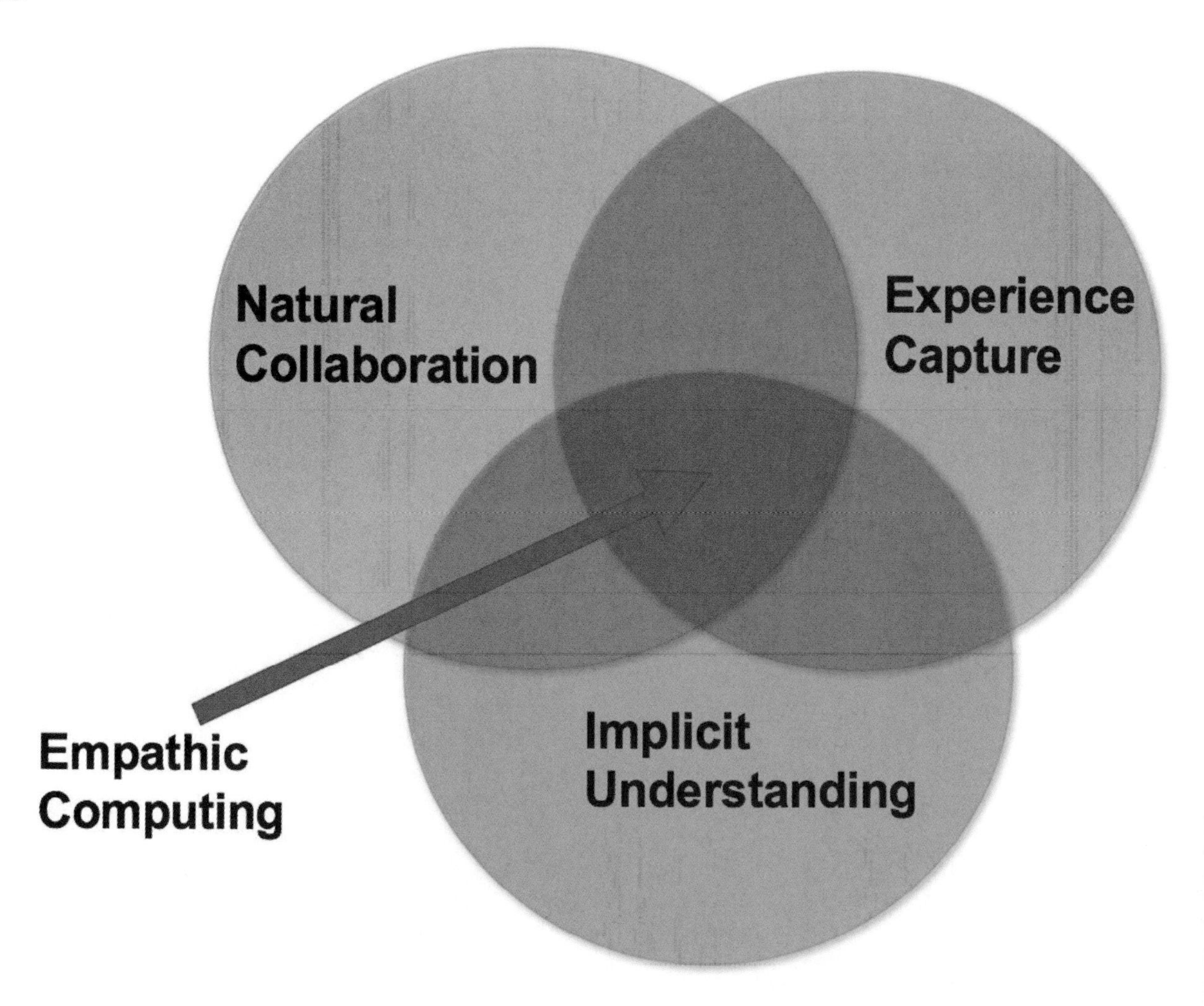

Mixed reality technology can be used to create radically different ways for people to communicate together. From the early days of virtual reality (VR), there was an interest in how the technology could be used for collaboration. In 1990, Jaron Lanier from VPL Research demonstrated Reality Built for Two (VB2)[1], a two user VR experience. This allowed two people in the same physical location to put on head mounted displays and entered into a shared virtual environment, playing games with each other.

More impressive, in 1995, the HIT Lab at the University of Washington demonstrated the Greenspace project[2] which created a transpacific shared virtual world where people in Seattle could share the same virtual space as colleagues in Japan. In Greenspace, up to four participants could enter a small virtual tea room in Japan and talk to one another. They each were given a virtual face and disembodied virtual hand to be able to support simple communication.

At the time the VR hardware cost over a million dollars and the communications bill for the few days the link was open was tens of thousands of dollars. Now, twenty years later, VR conferencing is becoming more commonplace. Companies such as High Fidelity, Sansar and Facebook are all developing collaborative VR spaces. AltspaceVR regularly has over 30,000 monthly active users of its shared VR platform, and collaborative VR applications like Rec Room and Big Screen are growing in popularity. This shows that there is a strong demand from VR users for shared virtual experiences.

One of the big attractions of using VR for conferencing is that it allows people to use some of the same communication cues they use in face to face meetings. Not only can they speak to one another, but users have virtual bodies that allow them to turn to face each other, shake hands, and make a rich range of non-verbal communication gestures. Finally, they can interact with the virtual environment around them, such as pointing to and talking about objects in the space, or playing games together. For example, in Rec Room, users can play table tennis, while people in Facebook Spaces can sketch together in 3D. This spatial collaboration creates a much higher degree of social presence than in traditional audio or video conferencing.

Collaborative Mixed Reality

VR conferencing has a lot of potential, and companies are investing many millions of dollars into collaborative VR platforms. However, mixed reality conferencing may have even more impact, because of its connection back to the real world.

Mixed reality (MR) is broadly defined as technology that mixes real and virtual worlds[3]. Unlike VR, which separates people from the real world, MR tries to enhance interaction in the real world or bring elements of the real world into VR environments. This means that MR technology can potentially be used for a wider range of applications compared to VR, and support more natural collaboration in the physical environment.

The benefits of using mixed reality for remote collaboration include[4]:

–Enabling people to get help from remote users on real world tasks.
–Bringing remote virtual people into a user's real space.
–Supporting transitions from shared AR to VR views.
–Using MR imagery to provide enhanced remote communication cues.
–Enabling users to share viewpoints and see through each other's eyes.
–Overcoming the seam between task space and communication space.
–Supporting natural spatial cues for remote collaboration.

Perhaps the biggest benefit is that MR conferencing is typically focused on sharing spaces, rather than sharing faces. For example, an MR system can use live streaming video to show the view of a user's workspace with a remote collaborator, not their face as in traditional videoconferencing. For many real world tasks, such as remote maintenance, it's more important to see what the person is working on, rather than their face when you're talking to them. Using this capability, there are a large number of domains where MR conferencing could be applied, from being used for remote expertise assistance in industry, to providing medical support in the operating theatre, and enhancing shared gaming, among others.

Early MR Collaboration Systems

Nearly 20 years ago, one of the first collaborative MR systems was an augmented reality (AR) conferencing application that placed live virtual video avatars of remote people in a user's real environment[5]. The user could turn over real name cards and see virtual faces appearing in front of them, showing live video feeds from remote collaborators. This was viewed in a head mounted display enabling the user to have a hands free collaborative experience. The main benefit was that it moved remote conferencing from a computer screen into the real world.

Researchers found that seeing a remote collaborator as a life size virtual video avatar in the real workspace provided a much higher degree of social presence, than seeing them on a computer screen. For example, one study comparing AR conferencing to a monitor based video conferencing system[6]. One of the people using the video conferencing system moved themselves very close to the monitor, assuming that this would help the remote person hear them better. However, when they tried the AR system and saw a life size virtual head facing them, they immediately moved back and gave them the same personal space as they would in a face to face conversation. This unconscious act was a strong indication of the increased feeling of social presence provided by the AR application.

WearCom[7] was a wearable AR conferencing application that allowed multiple virtual people to appear around a person using a wearable computer and head worn display. In this case, spatial audio was used to locate the voices of the people from their virtual avatars. Just as in face to face conversation in a crowded party, using spatial audio enabled people to easily discern between multiple speakers, even when they were saying almost the same thing.

Although this research showed a lot of promise, one problem was that the video avatars of the remote people were flat rectangles. If users looked at the people from the side, they would disappear, reducing the illusion that there were 3D virtual people in the user's space. This problem was solved a couple of years later with the 3D Live System[8], which used multiple cameras to capture people and enable them to be viewed from any angle. This created the illusion that a live 3D virtual person was standing in a user's real space. Once captured, the virtual copy of the real person could be streamed live into either AR or VR environments. This enabled the remote collaborators to use the same full body movement and gestures as in face to face collaboration.

The term mixed reality describes a continuum of interface technology from the purely real world to a fully immersive virtual environment. Most collaborative systems exist at discrete points on this continuum, such as face to face collaboration in the real world, or VR conferencing in an immersive space. However, MR technology can also enable people to transition along the MR continuum. For example, the MagicBook project was an interface that supported collaboration in a face to face, AR or VR view[9], or mixture of these. Using this system two people could read a normal book, but they could also look at the book pages through a handheld display and see AR content popping out of them. When a user sees an AR scene that is interesting, they can flick a switch on the handle and transition into an immersive VR experience. When in a VR scene, the user can look up and see their partner in the real world looking down at them as a giant virtual head in the sky. In this way, the MagicBook supported seamless transitions between shared real world, AR and VR experiences.

> Enabling people to **get help from remote users** on real world tasks.

These research prototypes showed that MR technology can be used to seamlessly place virtual collaborators into a user's real world, and provide unique collaborative experiences. Unlike VR conferencing, the MR interface enhances real world collaboration, and can enable a user to get help with real world tasks. Tests with these systems and others have found that people have more natural collaboration with an MR interface, and feel a much higher degree of social presence[6]. MR Conferencing can be much more like face to face collaboration than video conferencing.

Current MR Conferencing

The emergence of a new generation of AR and VR displays has led to the creation of new collaborative MR experiences. Microsoft with their HoloLens head mounted display has been showing an MR version of Skype. In this application, a user can position a virtual Skype window anywhere in space and see live video from a remote collaborator. At the same time, the HoloLens camera can stream live video to the remote collaborator so that they can see the local user's workspace and add AR annotations to it to help them complete real world tasks. In this way, the remote collaborator can feel like they are seeing through the eyes of the HoloLens user.

However, just like the AR conferencing application of nearly twenty years earlier, Skype on HoloLens places live video into a flat virtual rectangle. Microsoft's Holoportation project overcomes this by using multiple depth sensing camera to capture and live stream a 3D virtual model of the remote user into the local user's real world[10]. Combining this with HoloLens means that the local user can see a life size virtual copy of a remote user in their real world. Just like the earlier 3D Live system, Holoportation allows people to research how full body communication cues can be transmitted into remote real spaces.

There are also a number of startups beginning to work in this space. Mimesys provides one of the first MR meeting platforms, which can capture real people and bring them together into shared virtual spaces. Similarly, DoubleMe's Holoportal provides a lightweight version of Microsoft's Holoportation for shared meetings. These are both focused on capturing people, while Envisage AR is focusing on capturing a user's real surroundings and sharing it with remote collaborators. Over the next few years expect to see a lot more activity in this space.

These systems show that MR conferencing experiences can be delivered on current AR and VR commercial platforms. As the displays, tracking technology, capture systems and bandwidth continue to improve, the MR conferencing systems will continue to get better and better.

What's Next?

There are a number of developments occurring that will continue to improve MR conferencing and enable people to collaborate together more effectively than ever before. In particular, there are three important technology trends:

One of the big attractions of using VR for conferencing is that **it allows people to use some of the same communication cues they use in face to face meetings**

Natural Collaboration: As networking speeds increase it is possible to send more and more communication cues than ever before (i.e. video replaces audio only), leading to more natural collaboration.

Experience Capture: Technology is being developed that enables people to capture more of their experiences and surroundings than ever before, i.e. going from photography to 3D scene capture.

Implicit Understanding: Computers can now understand more about users and their surroundings than ever before. This allows them to recognise implicit behavior, such as where a person is looking.

At the junction of these three trends is a research area called empathic computing[11]. This is a research field that is focusing on developing systems that allow users to share what they are seeing, hearing and feeling with others. Unlike traditional conferencing tools, Empathic systems are designed to enable users to more deeply understand their collaborator's viewpoint, seeing through their eyes, hearing what they are hearing, and to some extent, knowing what they are feeling.

Empathy glasses[8] is one example of the type of MR collaborative experience that can be developed from an empathic computing perspective. This is an AR display that can also recognize face expressions and track eye gaze. When a person wears the glasses, a video of their workspace is sent to a remote collaborator as well as their facial expression and eye gaze information. The remote collaborator can see what the local user is seeing, but also knows exactly where they are looking. Further, they understand when they are feeling confused or unhappy about the task they are doing. This is one of the first wearable collaborative systems that shares eye gaze in a head worn AR display.

Goodbye Phone...

In twenty years, using a mobile phone to connect with friends will feel as old fashioned as using a landline does today. By then, mixed reality technology will enable people to see virtual copies of their friends in the real world with them, or enable them to see through their friends eyes and help them with real world tasks.

The VR and AR systems of today show glimpses of what is possible with mixed reality. MR systems can share rich communication cues in the real world and enable remote people to collaborate in ways that was never before possible. ■

References

[1] Blanchard, C., Burgess, S., Harvill, Y., Lanier, J., Lasko, A., Oberman, M., and Teitel, M. (1990). Reality built for two: a virtual reality tool (Vol. 24, No. 2, pp. 35-36). ACM.

[2] Mandeville, J., Furness, T., Kawahata, M., Campbell, D., Danset, P., Dahl, A., ... and Schwartz, P. (1995, October). Greenspace: Creating a distributed virtual environment for global applications. In Proceedings of IEEE Networked Virtual Reality Workshop.

[3] Milgram, P., and Kishino, F. (1994). A taxonomy of mixed reality visual displays. IEICE TRANSACTIONS on Information and Systems, 77(12), 1321-1329.

[4] Szalavári, Z., Schmalstieg, D., Fuhrmann, A., and Gervautz, M. (1998). "Studierstube": An environment for collaboration in augmented reality. Virtual Reality, 3(1), 37–48.

[5] Kato, H., and Billinghurst, M. (1999). Marker tracking and hmd calibration for a video-based augmented reality conferencing system. In Augmented Reality, 1999.(IWAR'99) Proceedings. 2nd IEEE and ACM International Workshop on(pp. 85–94). IEEE.

[6] Billinghurst, M., and Kato, H. (2002). Collaborative augmented reality. Communications of the ACM, 45(7), 64–70.

[7] Billinghurst, M., Bowskill, J., Jessop, M., and Morphett, J. (1998, October). A wearable spatial conferencing space. In Wearable Computers, 1998. Digest of Papers. Second International Symposium on (pp. 76–83). IEEE.

[8] Prince, S., Cheok, A. D., Farbiz, F., Williamson, T., Johnson, N., Billinghurst, M., and Kato, H. (2002). 3d live: Real time captured content for mixed reality. In Mixed and Augmented Reality, 2002. ISMAR 2002. Proceedings. International Symposium on (pp. 7–317). IEEE.

[9] Billinghurst, M., Kato, H., and Poupyrev, I. (2001). The MagicBook: a transitional AR interface. Computers & Graphics, 25(5), 745–753.

[10] Orts-Escolano, S., Rhemann, C., Fanello, S., Chang, W., Kowdle, A., Degtyarev, Y., ... and Tankovich, V. (2016, October). Holoportation: Virtual 3d teleportation in real-time. In Proceedings of the 29th Annual Symposium on User Interface Software and Technology (pp. 741-754). ACM.

[11] Piumsomboon, T., Lee, Y., Lee, G. A., Dey, A., and Billinghurst, M. (2017, June). Empathic Mixed Reality: Sharing What You Feel and Interacting with What You See. In Ubiquitous Virtual Reality (ISUVR), 2017 International Symposium on (pp. 38-41). IEEE.

[12] Masai, K., Kunze, K., and Billinghurst, M. (2016, May). Empathy Glasses. In Proceedings of the 2016 CHI Conference Extended Abstracts on Human Factors in Computing Systems (pp. 1257–1263). ACM.

Computers can now understand more about users and their surroundings than ever before.

The Trillion Dollar 3D Telepresence GOLD MINE

by Charlie Fink

I decided to find out if what Microsoft says is true: Remote volumetric telepresence and collaboration can, and will be done sooner than people think. The technical hurdles are complex, but have potential solutions, all well worth undertaking, as volumetric telepresence will certainly be one of the killer apps of augmented and virtual reality.

Rewind. It took the personal computer roughly fifteen years to hit an inflection point and become a consumer product everyone had to have. At first, its killer app email, which most people first got at work, didn't seem so revolutionary. Hardly anyone outside the company was using it. The network effect, a phenomenon whereby a service becomes more valuable when more people use it, hadn't kicked in. New technology always penetrates the enterprise before the home. Once people started getting Internet online services with a personal email address, it made the PC something everyone had to have at home. The telephone is another great example. The more people who got one, the more people had to have one.

Similarly, messaging and social media are the killer apps of smartphones. Our need to connect with other people follows us, no matter where technology takes us. New technology succeeds when it makes what we are already doing better, cheaper, and faster. It naturally follows that telepresence should likewise be one of the killer apps for both AR and VR. A video of Microsoft Research's 2016 Holoportation experiment suggests Microsoft must have been working on this internally for some time. Maybe even before the launch of the HoloLens itself.

Telepresence, meaning to be electronically present elsewhere, is not a new idea. As a result, the term describes a broad range of approaches to virtual presence. It breaks down into six main types:

Remote participants look like victims of a Star Trek transporter accident: Only 80% there.

Which one is a hologram?

A viewer looking through the red rectangle sees a ghost floating next to the table. The illusion is produced by a large piece of glass, Plexiglas or plastic film situated at an angle between viewer and scene. The glass reflects a room hidden from the viewer (left), sometimes called a "blue room", that is built as mirror image of the scene.

2D Video Conference Systems: These have become incredibly sophisticated and include eye tracking to help create presence for colleagues who are still seen on a monitor. Cisco's Spark System dominates the billion-dollar teleconferencing industry.

Robotic Telepresence: Describes any remotely operated vehicle with a driver's view such as remote underwater vehicles (ROVs) and unmanned aerial vehicles (UAVs), or RPAs (remotely piloted aircraft). NASA has long dreamed of true, real-time robotic telepresence, which was, in fact, one of the initial purposes of their VR research in the 80s. However, due to the time delay lag of signals to travel from Earth to Mars and back, NASA scientists can't directly tele-operate a robotic explorer like the Curiosity Mars rover in real time. However, it's possible astronauts aboard a spacecraft orbiting Mars would be able to.

Remote Experts: They use AR to see what you're seeing, although they cannot see you. They can even draw on the live feed you are sharing with them, interacting with real objects in your field of view in real time. Remote experts turn low skilled employees into highly skilled ones.

VR Telepresence: This allows us to share a virtual world like Oculus Rooms or AltSpaceVR where we are represented by an avatar. Today most avatars are cartoonish, but they will soon be able to use 3D volumetric captures taken on a cellphone to skin avatars that are eerily accurate. Lip sync (more precisely real time lip animation) and eye contact introduced by Sansar and High Fidelity, already can make you feel very, very present.

Our need to connect with other people follows us, **no matter where technology takes us.**

AR Telepresence: This allows two or more remote people to have volumetric presence in the same room, which Microsoft calls Holoportation, because it uses their HoloLens. This has been convincingly demonstrated, and now companies are seeking to bring that technology to business conferencing. However, not all the technical and practical issues around this have been solved. Several companies are working on solutions that could disrupt the teleconferencing business Cisco dominates. Cisco itself recently added a VR collaboration feature to Spark.

True Holographic Telepresence: This unaided volumetric holographic (visible to the unaided eye) presence can be done today with holographic projection, mirrors, and an invisible projection surface. This works well under very specific circumstances. In no way would the participants perceive each other, but to people outside the simulation, it is completely real. They'd see two (or more) people in remote locations in real life, interacting, on stage, without headsets, in a shared 3D space. However, the players could not see each other, they'd be looking past the reflections at a monitor. From the audience, you'd never know.

The Steven Spielberg movie *Minority Report*, also features augmented reality in the scene where data floats in front of Tom Cruise without a projection surface, visible to the naked eye, and he manipulates it with his hands. This would only be possible if Cruise's character had either contacts or some sort of neural input that could send images directly to his brain. Otherwise, projected holograms can only be visible to the naked eye if there is a transparent projection surface.

The HoloLens and other AR HMDs are equipped with inside out cameras. In order to create a telepresence app, however, an outside in camera that can face you and take videos of you is necessary.

I visited Steve McNelley, co-founder, and CEO of DVE Telepresence, in his workshop. DVE has been working for the Department of Energy and some of the largest companies in the world to provide what he calls the "only true telepresence". This requires three things, he explained, "absolute photorealism, perfect camera alignment for eye contact, and augmented reality images (holograms) appearing in space with no glasses required." DVE has a podium based system called "the 4D Telepresence Podium" which accomplishes all these things in a portable solution. The speaker behind the podium is captured in a remote location (such as a classroom room or a personal office) and projected in real time onto an invisible translucent surface and seen in the middle of the room by an audience. The speaker is projected onto the surface, and the camera is positioned to maintain eye contact with the audience.

Telepresence will happen very slowly, **and then all at once.**

DVE has demonstrated and patented many different technologies to create this holographic experience from OLED, LED, direct projection and a variant of an illusion enabling natural telepresence called "Pepper's Ghost," first demonstrated by stage artist John Henry Pepper in 1862. This method creates a "ghost" by reflecting an object onto a translucent surface, like a pane of glass, so the image seems to float in front of us. Today, DVE has advanced this to create bright solid looking people that look like they are really in the room, as can be seen in the above image where I appear to be in the same room as a hologram with Zach McNelley, DVE's 3D content creator. The two requirements are a perfectly black background and a translucent projection surface.

"Pepper's Ghost" was most famously deployed in Disneyland's Haunted Mansion to create the illusions of spectral dinner parties and hitchhiking ghosts. In fact, the *Star Wars* Jedi Challenge VR Headset from Lenovo uses a similar method of bouncing an image off a mirror onto a transparent

Valorem's HoloBeam brings volumetric conferencing to life using standard Internet with normal bandwidth requirements without latency. Note the digital "dust" around the figure. The system today requires dedicated computers to encode and decode both ends of the conference in real time. Photo courtesy Valorem

projection surface to create the illusion of 3D characters floating in space before us.

Microsoft has been promoting another vision of telepresence and remote collaboration for the HoloLens that they call Holoportation which allows participants in remote locations (they were actually down the hall) to be present in each other's physical reality. Multiple 3D cameras were placed in each room. These inputs were fed into local computers which broadcast the compressed 3D image to the user's HoloLens. This video was posted on November 2016, which means that MS engineers must have already been working on Holoportation before the HoloLens was released in March 2016.

Microsoft Research's Room2Room is a life-size telepresence system that uses projected augmented reality to enable co-present interaction between two remote participants without using a HoloLens. This solution recreates the experience of a face to face conversation by performing 3D capture of the local user with 3D cameras and then projecting the volumetric copy into the remote space at life size scale, instead of using the HoloLens. This creates an illusion of the remote person's physical presence in the local space, as well as a shared understanding of verbal and non-verbal cues (i.e., gaze, pointing) as if they were there. This is what happens when there is no mediating projection surface, which for research purposes, the engineers have eschewed in favor of flexibility and intelligence.

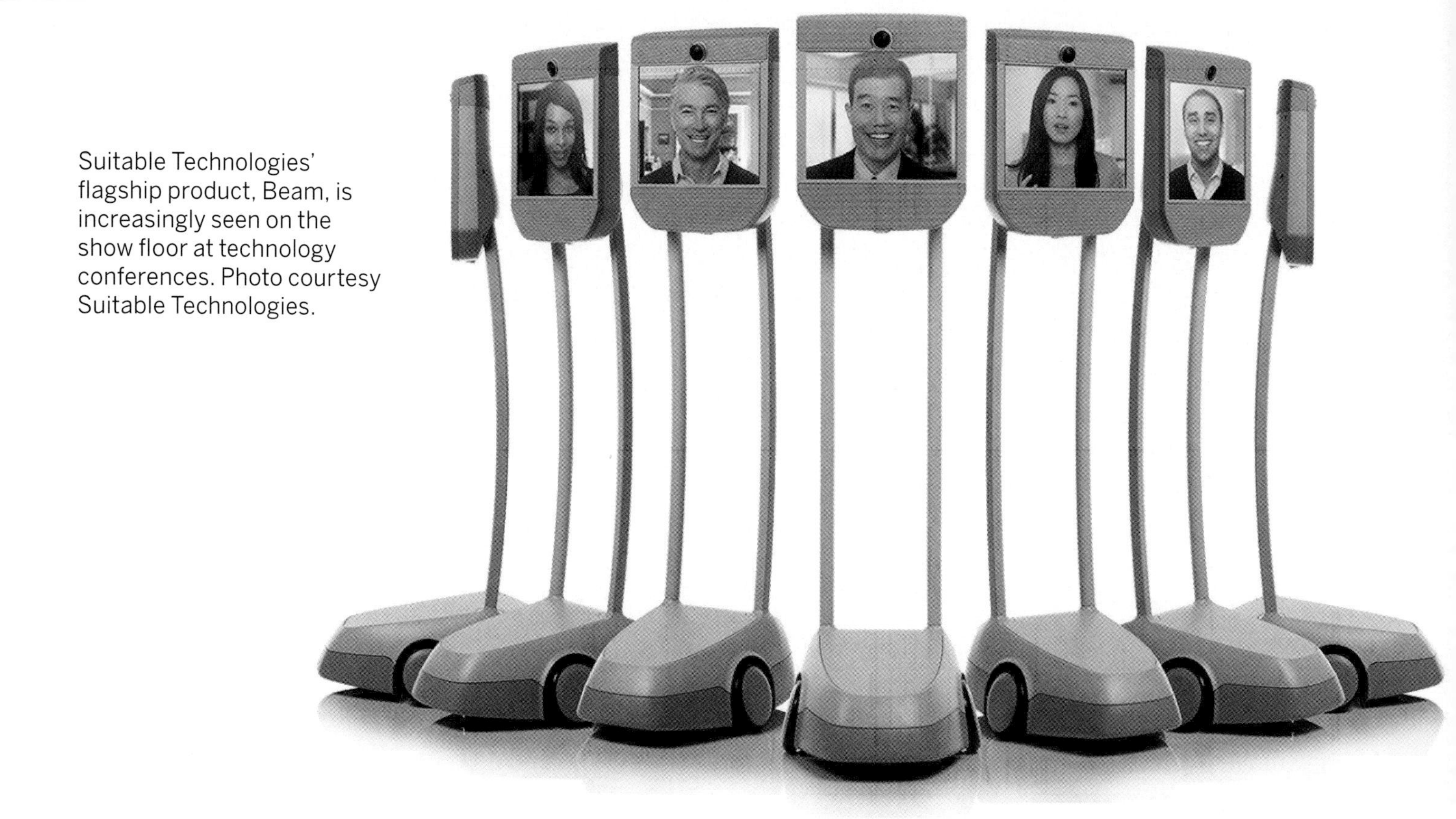

Suitable Technologies' flagship product, Beam, is increasingly seen on the show floor at technology conferences. Photo courtesy Suitable Technologies.

In early 2017, Microsoft spent what appears to be millions of dollars to create a video which portrays the future (or one potential future) of holographic telepresence, called *Penny Walks in a.k.a. Envisioning the Future with the HoloLens.* "Penny" is an extraordinarily well produced science fiction video dramatization of a telepresence use case, starring a retail designer (Penny) and her client in Asia. There's more than just telepresence going on. The client also has a floating, visible, seemingly sentient digital assistant, a sort of evolved Clippy. That aside, the subtle but ambitious scale of the simulated use case in the demo isn't crazy, far off, or impossible. But, remember the network effect. It needs scale to reach that magic inflection point, where rooms are scanned in real time by 3D cameras, awaiting Penny and the rest of us.

Microsoft's research teams continue to explore Holoportation, along with several universities. Notably Warsaw University of Technology in Poland, where Marek Kowalski and Jacek Naruniec have been developing a Holopresence app, LiveScan3D.

LiveScan3D does real time 3D reconstruction by using multiple Kinect V2 depth sensors simultaneously to produce a colored point cloud, compressing the 3D video inputs. Each Kinect V2 sensor is connected to a separate computer. Each of those computers is connected to a server which allows the user to perform calibration, filtering, synchronized frame capture, and to visualize the acquired point cloud live in a remote location. Consistent with their role as academics, Kowalski and Naruniec have shared LiveScan3D as open source code on Github, allowing others to build on their work.

Private companies are also making impressive progress with 3D volumetric conferencing using both VR and AR. Notably, Valorem, whose system enables multiple participants in Europe, India, and North America to be volumetrically present in the users' physical office in real time. Mimesys and Meetingroom.io use VR to create volumetric presence in a shared virtual world.

René Schulte, who is leading the HoloBeam development effort for Valorem, is based in Dresden, Germany. He described how the company's unique 3D real time conferencing system works, and how it is transforming collaboration among his cross continental teams in Germany, Seattle, and India.

"This was captured real time in HD using a depth camera to collect 3D volumetric video point cloud data consisting of color and depth information. The point cloud data is then streamed or 'beamed' across the Internet over a customized WebRTC stream. The holographic stream is decoded by an app and rendered in real time 3D, providing a shockingly good volumetric representation of the senders' likeness on VR and mixed reality devices like the HoloLens, but also other devices are enabled by our cross platform development approach. It runs over a normal internet connection and requires 3-5 Mbits/sec bitrate, and even works below 1 Mbit thanks to our adaptive, depth encoding and streaming (adaptive streaming is what Netflix does to adapt to your connection speed). It's real time without delay, and even works if the parties are behind firewalls, for example, in corporate network settings. There's no special connection or setup needed. The connection is established via a routing mechanism to connect peer-to-peer for the best transfer rates."

The HoloBeam system does not provide the kind of resolution we saw in the MS Holoportation videos, but we're now told those 2016 videos were only local proof of concepts, not something to set up in remote locations. In contrast, Valorem's system today produces 3D volumetric video with a simple setup, using off the shelf hardware.

The system can have varying amounts of "dust artifacts" (drop out), depending on how much the adaptive streaming has to ratchet down the bandwidth. As a result, remote participants look like victims of a *Star Trek* transporter accident: Only 80% there. However, everyone I talked to, and everything I experienced myself researching this story has proven that 80% is enough to create deep, compelling presence.

The holographic point cloud will have more resolution in the future with improvements, not only with increasing depth camera resolutions and bandwidth, but algorithms that fill in missing pixels in decompressed video files to reduce the broadcast dropout or dust as the Holoportation products evolve.

Schulte sees bright things on the horizon. In the office of the future, multiple depth sensing cameras could literally merge it with remote locations around the world, or we could even just use our mobile phones which start to integrate depth mapping sensors and dual lenses in consumer products. The next guy knocking on your door could literally be in China. Valorem expects to start broader trials with clients in early this year.

Mimesys of Paris, and Meetingroom.io, of Dublin, are startups taking a different approach, using VR as the basis for shared collaborative meeting spaces, which can include users on multiple devices like PCs and smartphones. At their core, these systems bring volumetric captures of remote participants into a virtual room much like we see in social VR like AltSpaceVR and Oculus Rooms today. Mimesys allows users to log into their virtual meetings using any device, including HoloLens, tablets and smartphones.

Mimesys Connect is centered on a shared virtual conference room that allows participants to import and share 3D objects, watch videos, and do just about everything you can do in a real business meeting. Unlike the social space for consumers, however, the participants are not avatars, but volumetrically present. This distinction is incredibly important.

The experience I had using Mimesys Connect on the Vive with the founder and CEO, Remi Rousseau, was extraordinary. We saw one another's volumetrically captured 3D avatars, wearing our headsets, and were able to share and pass objects in a virtual conference room. The feeling of being present with him was very strong.

Rousseau believes the VR-centric approach is the most flexible and easy to use. "HoloLens Teleportation doesn't allow users to share and collaborate the way they do with Mimesys Connect. We can't collaborate on a shared whiteboard, for example." I asked Russeau about barriers to entry and how his small startup, in use with perhaps a dozen pilot clients, could defend this kind of VR approach from low or no cost competitors like AltSpaceVR and Oculus Rooms.

"There is a potential risk, especially regarding Facebook spaces," he said, "which is also why we focus today on B2B rather on B2C. That being said, the communication space is huge. Platforms like WhatsApp, Messenger, Hangouts, Facetime, co-exist today, and that would probably be the same for VR and AR communication. There will be different experiences with different audiences."

"We're still at the beginning, but the portability is the game changer here," added Jonny Cosgrove, founder and CEO of Meetingroom.io. "C-suite and sales directors can meet and manage salesforces. Companies can engage with more customers."

OJ Winge, currently SVP of Cisco's Video Technologies Group has been working with telepresence in one form or another for most of his career. "Cisco's Spark system already provides a new richness of experience," he said. "Right now quality isn't good enough for volumetric telepresence, which we see as something different from Spark. Complementary. Different. But not a replacement. For a normal meeting, the technology needs to be transparent and natural." He is confident of Spark's position and plans to grow the business.

Telepresence will happen very slowly, and then all at once. Dramatically disrupting not only the conferencing business, but business management and collaboration itself. ■

Who Can Build The AR Cloud?

By Ori Inbar

The AR Cloud is such a mega project that perhaps only three companies in the world have deep enough pockets and sufficiently big enough ambitions to tackle this.

Apple, Google, and Microsoft could each some day pull off the full AR Cloud. Actually, if they don't they, will get screwed by whoever does.

But, if history teaches us anything, it's that it's going to take a crazy startup to put together the building blocks and prove the concept. Because only a startup can have the audacity to believe it can build it. It'll take a crazy good startup.

Since it's too big for one startup, we may see different startups tackling different parts of it. At least at first.

Apple Apple's CEO Tim Cook has been touting AR "Big and Profound" and Apple seems to have big ambitions with AR. ARKit was a nice low hanging fruit for Apple. They had it running in house (thanks to the Flyby acquisition) and had in house expertise to productize it (thanks to the Metaio acquisition). However, because Apple is unready to take on the creation of a full blown AR Cloud, the potential of ARKit has not yet been realized. It might be due to privacy and security concerns around the data. Or perhaps they're hiding it well, as they tend to do.

Google The AR Cloud is Google's natural next step in "organizing the world's information". Google probably possesses more pieces of the puzzle to deliver on this mega project than anyone else, but i'm not holding my breath.

Tango launched as a skunk works project in 2013 with a vision that includes three elements:

Motion Tracking: Using visual features and accelerometers / gyroscope).This was spun out as ARCore.

Area Learning: Gives the device the ability to see and remember the key visual features of a physical space (the edges, corners, other unique features) so it can recognize that area again later, and storing environment data in a map that can be re-used later, shared with other Tango devices, and enhanced with metadata such as notes, instructions, or points of interest.

This sounds pretty close to the AR Cloud vision, but in reality, Tango currently stores the environment data on the device, but doesn't have the mechanism to share it with other Tango devices or users. The Tango APIs do not natively support data sharing in the cloud.

Depth Perception: Detecting distances, sizes, and surfaces in the environment. This requires the Tango cameras.

Google announced at this year's I/O conference a new service that points to another piece of the AR Cloud dubbed the Visual Positioning Service (VPS). However, based on developers inquiries, it's fair to classify it as "very early stage".

Of course Google has already built a pretty elaborate 3D map (for outdoors) thanks to its street view cars. Google may have all the AR Cloud pieces together within a few years. But Google API tends to be limited and may not be good enough for an optimal experience.

So Google is a natural top contender to own the AR Cloud, but unlikely to be the best or even the first.

Microsoft
Microsoft's CEO revitalized focus has been clearly communicated on the cloud and on holographic computing. With the HoloLens, the most complete glasses system to date (though still bulky and pricey), Microsoft has delivered a fantastic capability to create point clouds that persist locally on the device. However, it's not capable to share it natively across multiple devices. Since Microsoft had no good hand to play in the current mobile AR battle, it skipped a generation straight to smart glasses and is leading the pack. But because of the anticipated slow adoption of smart glasses, their AR Cloud may not see much demand for a while.

Facebook, Snap and perhaps Amazon, Alibaba and Tencent may have that ambition as well but are years behind. A great opportunity for some of the startups above to get acquired.

Other contenders include Tesla, Uber, and other autonomous cars and robotics companies which are already building the 3D map of the outdoors or indoors real world, but without necessarily a special focus on AR.

Crazy good startups who have tackled parts of the AR Cloud and been acquired:

13th Lab (Facebook)
Obvious Engineering/Seene (Snap)
Cimagine (Snap)
Ogmento->Flyby (Apple). Flyby also licensed its tech to Google.
Georg Klein* (Microsoft)

Startups currently tackling the AR Cloud or parts of it:

YOUar (USA)
Scape (UK)
Escher Reality (USA)
Aireal (USA)
Sturfee (USA)
Occipital (USA)
Fantasmo (USA)
Insider Navigation (Austria)
InfinityAR (Israel)
Augmented Pixels (Ukraine)
Kudan (UK)
DottyAR (Australia)
Meta (USA)
DAQRI (USA)
Wikitude (Austria)
6D.ai (USA)
Postar.io (USA)
ContextGrid (USA)
Vertical.io (USA)
MAP (USA)
ReScan [country?]
Synth (Australia)
<Your Startup Here>
What are the big players doing about it?

** Georg is a person, not a startup. But has single handedly outdone many companies. See his first ARKit-like demo on the iPhone 3G in 2009!*

The Power of The Blockchain

Philip Rosedale Nails Blockchain in 500 Words
Blockchain and VR/AR...Wait...What?!
The Coming Intersection Between Blockchain and VR/AR

Interview Transcript: Philip Rosedale with Charlie Fink and Robert Fine 10/18/17

Philip Rosedale: If Blockchain were around when we started Second Life we would have built everything around it, and indeed that is some of the work we're undertaking now [with High Fidelity]. I think that Blockchain is super important for money, for digital assets, and most importantly for identity. Not just in the virtual world, but in the real world.

Virtual things are only as real as they are durable. You want a couple of things to be true. Ideally, you'd like virtual things that you buy, like a virtual Ferrari, to be persistent and real even if the company that sold them to you disappears. You'd like there to be the strongest possible statement about the ownership of that object, made in as broad a context as possible. The real world gives us atoms to do this. We need to come as close to that as we can. The idea of a blockchain is very, very powerful in that regard. When I sell you something like that Ferrari, I can write to the blockchain and say, "Charlie owns this Ferrari." Once I've done that, I can't take it back. I can't say he doesn't own it. I can't edit history. I can't censor Charlie. I can't take his Ferrari away.

And, moreover, there's a distinction made between this general statement that he owns this thing called a Ferrari and the actual pixels that represent it. Those can be drawn in different ways at different times in different worlds, but the Ferrari, minimally, can be this matter of public record. It can be a thing you enter into a public ledger that proves that you own it. I think in virtual worlds there has to be a simple system that allows you to say, "This digital thing is mine." The other thing the blockchain does is it stores that information in a way that only the owner who has the private key can be associated with that little chunk of the blockchain. The owner can update the object, sell it, move it around. No company or central organization is able to do that once it's printed in the blockchain.

Right now, do we feel a Pandora station or an iTunes song is more durable? Because you've downloaded the iTunes song. You have that song. It can't just be taken away from you. Where, if you're using something like Pandora or Rhapsody, the deal might break down between the Beatles and those hosting companies, and you might not be able to listen to the Beatles anymore. So I think that's the power of the Blockchain. ■

The REAL WORLD gives us ATOMS to do this

VR AND AR STRATEGIES OF THE WORLD'S BIGGEST Technology Companies

By Charlie Fink

Tim Merel of Digi-Capital says it's all about what the guys at the top of the pyramid do, and he's right. Except, when they are severely disrupted by technology. The way Google disrupted online services like AOL. In this case, all eyes are on Magic Leap, a private technology firm working on an AR HMD that has received nearly $2 billion of funding to date from Google, Andreessen Horowitz, Alibaba, Warner Bros., Disney, Qualcomm, Kleiner Perkins, Vulcan, and a raft of well known finance players like J.P Morgan. They are simultaneously building the technology and a unique, proprietary operating system for it. If they succeed, Magic Leap could be another Apple. After all, none of the world's leading technology companies was born before 1975. During periods of profound technological disruption small companies can become the largest in the world. Facebook is only thirteen years old. Google, twenty.

However, one cannot truly understand what's going on with VR and AR without understanding the strategic initiatives of the largest technology companies in the world, and as a writer it is not easy to see inside secretive large companies. You end up reporting on reporting.

Keep in mind, big companies don't own the answers. And they are blinded by who they are, and constrained by their cultures. Still, the all-time home run leader gets as many trips to the plate as he needs to hit the ball out of the park. So it is with Google. So what if Daydream fails? They can acquire the winner, as Facebook has consistently done. Sometimes for what seems to be astronomical sums of money. If Rift completely tanked, even after a five billion dollar investment, how much would it hurt Facebook? This is why today's large companies like Google and Intel maintain well funded active venture groups that seed many potential companies, giving themselves the inside track on technology, talent, partnerships and potential acquisitions.

Apple's enormous install base of six hundred million iPhones gives it the power to instantly put new technology into the market, as they did in September 2017 when they released ARKit as part of iOS 11. Although there are nearly two billion phones using the Android operating system, Google does not control those handsets or those users. It is much more reliant on the three year phone upgrade cycle than Apple. Three years can be a very long time in consumer technology today. This is the reasoning behind Google's acquisition of HTC's smartphone business.

Let's take a high level look at some of the big technology companies and their strategy for VR and AR.

Microsoft (NASDAQ: MSFT, market cap $650 billion)

In early 2016, Microsoft introduced HoloLens, an untethered AR HMD with a narrow 30 degree field of view, priced at $3,000, primarily for developers building enterprise applications. In November 2017, Microsoft introduced Windows 10 MR and a range of fully occluded VR HMDs from Acer, Lenovo, HP, Dell and others, joining PSVR, HTC Vive and Oculus Rift as the fourth high-end home system. Without external tracking lighthouses, the Windows 10 MR VR headsets have inside out cameras instead of lighthouses that track the controllers. Despite its confusing name, Windows 10 MR VR brings high-end VR to more computers, as users don't necessarily need a high-end graphics card for every VR application.

Three years can be **a very long time in consumer technology** today.

Google (NASDAQ: GOOGL, market cap $725 billion)

In 2015, Google introduced the Cardboard VR viewer for mobile. Since then some 20 million have been sold or given away (they are $3 on eBay.) The Google mobile VR platform, Daydream, has been hampered by the lack of compatible phones, making 2017 a disappointing debut year. Stand alone (untethered) VR headsets like the forthcoming HTC Vive Focus and an unspecified compatible HMD from Lenovo have failed to materialize. HTC has decided to launch the Focus in China, using its own Viveport distribution platform. The phone upgrade cycle will bring many new users to the well reviewed Pixel 2, but now that Google has acquired HTC's handset manufacturing business, there should be many more Daydream compatible phones coming to market in 2018. Google also has a substantial investment in Magic Leap (see sidebar).

Analysts caution it may be more like having **an iWatch on your face.**

and Vive, as well as on mobile devices. Amazon also plans to make its tools compatible with Google's Pixel phones. Like IKEA, Amazon also offers 3D models to users in its home category so that consumers can visualize how a product will look in their home.

Amazon is acquisition minded, and not just in the digital world. However, it would not be surprising to see the retail giant doing a lot more in the area as it begins to scale. How shopping will work in a fully immersive virtual world has not been figured out, but Amazon will certainly be among the first to do it.

Baidu (NASDAQ: BIDU, market cap $81 billion)

Baidu has made considerable domestic investments in VR/AR. Through its online video platform iQIYI.com, Baidu has its own exclusive platform for Chinese VR video content. They also launched a partner incentive program that will allow developers to make VR content around copyrighted movies, shows and games. Baidu is further investing in non-domestic companies such as VR/AR company 8i in LA. Financial services is also a growth area for Baidu, and the company needs to expand in this segment to stay competitive with players such as Alibaba and Tencent. Baidu is therefore expected by many to invest in fintech VR.

Alibaba (NYSE: BABA, market cap $455 billion

Alibaba has invested in and built VR shopping experiences for its 400 million users. Merchants will be able to create their own VR enabled shopping experiences. Alibaba´s e-commerce business is also working with Alibaba Pictures and Alibaba Music to develop VR content. In addition, Alibaba's finance arm is working on a technology that could potentially allow users to shop inside virtual stores without having to take off their virtual reality goggles to settle payments.

Tencent (TCEHY:US, market cap $463 billion)

Tencent pushes content development as well, and has made deals with both local as well as non-local content producers to create content for the company´s production studio, Tencent Pictures. Tencent has focused much of their investments in the games market, and on the VR side they have invested in Epic Games and AltspaceVR (recently sold to Microsoft). However, Tencent has also invested in Meta, a well funded Silicon Valley startup that makes AR headsets for the enterprise market. Tencent, like Baidu, has also invested in fintech VR.

Conclusion

There are many smaller public companies that are household names whose activities can still make markets. Particularly Snap, which is well positioned to leverage mobile AR. Snap Spectacles, though fashionable and cleverly marketed, only took short videos for use exclusively with Snapchat. HTC, having sold its handset business, is focusing on VR, and is now releasing a potentially game changing untethered all-in-one that is plug and play. ODG and Vuzix make affordable, wearable AR headsets for the enterprise and are also developing consumer versions. The ODG R-8 is now on the market in China.

While the long term vision of the industry is on head worn mobile computing, the near term action will be on the smartphone. Improvements in infrastructure will make mobile AR increasingly useful, and is likely to stimulate demand for HMDs in the long term. ■

The author would like to thank Dr. Annika Steiber for contributing research about Baidu, Alibaba and Tencent for this chapter.

Magic Leap

Magic Leap (ML) is a secretive startup based in Ft. Lauderdale, Florida, founded by Rony Abovitz, an inventor of medical devices who had made several hundred million dollars selling his company. In 2011, he pivoted his post wealth media company toward creating a new entertainment ecosystem, building his own operating system and hardware for a lightweight, fashionable AR headset with a wide field of view. If it can do this, Magic Leap will instantly become one of the most important companies in the world. This is why it has raised almost $2 billion dollars at a $4.5 billion valuation from the biggest technology companies (Google and Qualcomm), media companies (Warner Bros., Disney), the e-commerce giant Alibaba, the biggest names in venture capital (Andreessen Horowitz and Kleiner Perkins), and financial institutions (JP Morgan and Fidelity). In terms of venture activity in AR, this is an incredible raise. They are said to be burning through $100 million a month. On December 20th, 2017, Magic Leap finally gave the world it's first look at their augmented reality glasses, called the Magic Leap One: Creator Edition (ML1), which it expects to release sometime in 2018. They are going for the HoloLens audience: Developers, prosumers, agencies, etc., who are able to pay a premium price for a computer they will mostly have to program themselves.

Brian Crecente of Rolling Stone went to the secretive company's Florida headquarters for an exclusive first look. The photos ML concurrently released reveal a sort of steampunk HoloLens attached to a palm sized CPU, or what the company calls its "Lightpack". Crecente reports the ML1 has a somewhat wider field of view than the 30° of the HoloLens, but it is still not great.

ML has the ability to detect the space you are in with artificial intelligence, and to place objects that remain persistently anchored in space. Meaning, when you return to that location the next day, screens, post-its and other user created content will still be there, anchored where you left it. ML's

proprietary chip parses physical reality into cues for the brain, and bounces those cues through a light field optical display, bouncing light directly onto the retina, and giving virtual objects the same focal properties as physical objects, thus creating a powerful illusion and sense of presence.

Given the expected price of the ML1 and the limited amount of software offered with the developer edition, it's unlikely consumers will be breaking the doors down. "I would say we are more of a premium computing system. We are more of a premium artisanal computer," Abovitz told Crecente. It will take until 2019 for an ecosystem to develop that would support a mass market of AR glasses. Like Oculus and HTC in virtual reality, ML is going to have to support early developers who will have a hard time recouping their development investment before an app market comes to fruition. And we still don't know much about the ML1. How will it interact with the physical world without critical infrastructure like the AR Cloud, which we discussed in Chapter 10. Meanwhile, another company, Avegant, which used to make wearable media viewers, is also developing a light field technology it hopes to sell to Magic Leap's competitors.

In a late night tweetstorm at the end of November 2017, Abovitz said "We are not chasing perfection — we are chasing 'feels good, feels right.' Tuning for everyday magic. It is like tuning a guitar — when it sounds good, and it plays well, done. The feedback loop is everything. We are quietly learning with early developers - and listening to what they want and need."

WILL CHINA be the winner IN VR/AR?

By Dr. Annika Steiber

Dr. Annika Steiber is an international authority in the field of management and innovation, lecturer at Santa Clara University, and the author of several well known management books: The Google Model, The Silicon Valley Model-Management for Entrepreneurship, and most recently Management for the Digital Age: Will China Surpass Silicon Valley?

When new technologies and markets are being developed, they generate intense interest in two questions: "What forms will these things take?" and "Who will be the leaders?" And in recent years, in just about every field, that second question leads to a third: "What are the Chinese doing?"

China's emergence as a global force in innovation is becoming one of the biggest business stories of the 21st century. For example, one measure of innovative activity is the filing of international applications for patents under the global Patent Cooperation Treaty (PCT). China ranked third on that measure in 2016, closing in rapidly on the U.S. and second ranked Japan. The World Intellectual Property Organization noted that China "has posted double digit growth [in PCT filings] each year since 2002." Two Chinese companies, ZTE and Huawei, topped the list of filings by firms; and WIPO's Director General commented that these trends reflect the country's "journey from 'Made in China' to 'Created in China.'"

There are skeptics, of course. They may point out that patenting activity doesn't necessarily result in actual new goods brought to market. They may also point out that many of China's innovative high growth companies have grown primarily by serving their own domestic market. Internet giants like Alibaba, Baidu, and Tencent are household names within China, where they have immense user bases, yet they aren't nearly as well known elsewhere. And while it remains common for the rest of the world to use physical products manufactured in China, products invented and developed in China are, with a few exceptions, much less common.

All of that appears to be changing, however. As this chapter will show, China's capabilities for innovation have been growing tremendously, in many ways. Chinese companies already are recognized as world class players or even world leaders in diverse areas that include sharing economy services, electronic payment and finance coupled with e-commerce, consumer drones, electric cars, and more.

Speaking directly to the subject of this book: Both Chinese companies and the Chinese government have launched concerted efforts to become leaders in virtual and augmented reality. Moreover, in VR/AR, as in other fields, they are able to deploy strengths and advantages that have been little discussed thus far.

The rest of the chapter is in three main parts.

First, a general overview of China's growth as an "innovation country."

Next, a closer look at Chinese activity in VR/AR.

The final section explores a topic that is the focus of this author's research—and that may be an important factor to consider, in deciding whether Chinese companies can become leaders in VR/AR: Some have adopted new management models explicitly geared to innovation, flexibility, and rapid growth.

China's Rapid Learning Curve To Innovation

The Chinese economy has grown spectacularly since 1980. Per capita GDP in real US dollars rose from US$348 in 1980 to an estimated US$6,894 in July 2017. During the same period, according to World Bank figures, the number of Chinese people living in extreme poverty was reduced by about 800 million, with many rising from low wage status into the middle class or higher.

Conventional wisdom says these results are due to economic reforms that unleashed the power of the free market by allowing private businesses to start and grow. But while free market forces have been a vital factor, there is more to the story.

China's central government (and the provincial governments) have implemented an ongoing series of reforms, along with various official plans and strategies. By doing so, they've arrived at a powerful combination. It's one that mixes potent features of a market economy with those of a planned economy.

Companies in the private sector compete to develop products, services, and business methods that meet buyers' needs. Meanwhile, the government sector takes responsibility for setting and supporting the country's long term vision. And, especially over the past two decades, government efforts have focused on increasing China's innovation capacity.

The Government's Roles

Key policy measures and strategic thrusts by the government have been numerous. For simplicity, most could be put into three broad categories: Building up the innovation infrastructure, to provide settings or "platforms" for activities such as R&D. Identifying and supporting promising new fields (including by direct investment). And building up human capital through the higher education system. Here are just a few examples of each.

Innovation infrastructure: Early in the reform period, China's government began promoting the formation of national science and technology industrial parks (STIPs). By 2006 there were 54 national STIPs housing over 43,000 high tech firms. And while the number of STIPs has stayed fairly constant since then, other types of innovation hubs continue to grow. The town of Shenzhen was the first location in China to be designated a Special Economic Zone (SEZ), with favorable policies for opening up the region to foreign trade and investment. Shenzhen is now a burgeoning major city, ranked by The Economist as one of the world's innovation hot spots.

Alibaba processed **US $17.8 billion in online transactions** within 24 hours, more than Brazil's total e-commerce volume for the entire year.

The "opening up" policies expanded country wide, and according to the business researchers George S. Yip and Bruce McKern, China in 2016 had more than 1,500 R&D centers established by foreign firms. (The firms include GE, Oracle, Daimler AG, Medtronic/Covidien, Toyota and many others.) Meanwhile, Chinese companies have increased their own spending on R&D, hitting record combined figures in 2016, and, encouraged by the government, they have ramped up "outward" investment by acquiring foreign assets and opening R&D facilities in places from Silicon Valley to leading European innovation centers.

Identifying and supporting promising new industries. In 2006, the Chinese government issued its National Long Term Science and Technology Development Plan 2006-2020, which laid out a strategy to become an innovative country within 20 years. The goal was for China "not only to catch up with the West, but to re-establish itself at the forefront of technological innovation," as it once had been many centuries ago.

This Plan has been complemented by many other government plans and strategies, often targeted to particular fields. For example, the government's 2011-15 Five Year Plan included policies to promote e-commerce and electronic payments, in which China was already on the way to world status.

The strategies also include direct investment in new companies in growth industries such as IT. During 2016, a year when private venture capital investments in China hit an all-time high of over US$31 billion, the public sector stepped in to raise the bet dramatically. "Many provincial governments in China have also started to invest in companies, after Beijing identified entrepreneurship and innovation as the country's new growth engines," the South China Morning Post reported. According to Bloomberg, the Chinese provinces have dedicated a combined total of US$445 billion for VC investments. Such investments are seen as part of the price that must be paid to sustain China's economic growth, since older industries are slowing or even contracting, and new growth areas must be found.

Higher education: In 1999, the central government launched a program to massively expand university enrollment. According to figures reported by the World Economic Forum, eight million college students were scheduled to graduate in China in 2017: Ten times the number in 1997, and twice the number of 2017 college graduates in the U.S. Science and engineering studies have received particular attention. And given that China's education system is sometimes criticized as an impediment to innovation, with its emphasis on rote learning, it appears that quality has risen along with quantity.

The U.S. News & World Report (USN&WR) global rankings reflect this. China's Tsinghau University replaced MIT as the top rated engineering school in 2015; China and the U.S. now each have four such schools in the USN&WR top ten; and China now awards more PhDs in STEM fields than U.S. based universities do. Moreover, to help convert university work to practical innovation, the Chinese government provides grants to entrepreneurial students who have good plans for starting new companies.

That's a brief overview of government activities. Now consider that government support is just one part of what makes China a formidable competitor.

The Effects of China's Home Market

With a population base of 1.3 billion, not only is the Chinese domestic market becoming more free and prosperous, it's staggeringly big. This means intense competition, since a big market breeds plenty of market entrants. "Hundreds of thousands of new Chinese companies have made this country the world's most competitive business environment," according to the consulting firm Strategy&, which finds China to be the world's largest and fastest growing source of entrepreneurial startups. "By comparison, competition in the US is mild," wrote the US based technology reporter Clive Thompson.

But the competition drives innovation, and China's market size allows "winning" companies to earn huge revenues quickly. Two examples: In an online sale that introduced a new smartphone, the firm Xiaomi sold 800,000 of the phones in only 12 hours. And during its annual "Singles Day" promotion in 2016, e-commerce titan Alibaba processed US $17.8 billion in online transactions within 24 hours, more than Brazil's total e-commerce volume for the entire year.

Sales figures of this magnitude can give Chinese companies sizable sums to invest in exploring new lines of business. Furthermore, as one expert observer told the author in an interview, China's extremely large home market is "forgiving." That is, companies can experiment with new ideas in selected parts of the market without great risk to their overall brand reputation.

And there is still more to the picture. Chinese customers are avid buyers and users of information technology. For instance, they are heavy users of the Internet for shopping and other tasks, partly because China isn't blanketed by bricks-and-mortar chain stores, as other countries are, and partly because traffic jams and pollution in major cities make them want to do as much online as possible. During the decades of rising prosperity that lifted many Chinese beyond bare survival mode, they've also tended to "leap frog" directly to the newest and most advanced technologies. For instance, going directly to mobile IT devices and e-finance, instead of paper checking accounts.

These factors have created a large base of IT savvy users, which helps pave the way for acceptance of new offerings. And, while it's true that many Chinese companies have tailored their goods and services expressly to the home market, they are intent on entering global markets as well. One example here is Baidu, which leads the Chinese market for Internet search. The company's search engine is fine tuned to work well in the Chinese language, and the website carries news, ads, and entertainment content targeted to Chinese society. But Baidu is venturing into new areas that include AI, big data analysis, and self driving cars. These technologies have worldwide potential, as they could be used anywhere. The company is developing them at facilities both in China and overseas (Baidu has an AI Lab in Silicon Valley). And the self driving car venture has an array of foreign development partners.

From Copying to Innovation

In sum, the Chinese economy has indeed gone through a rapid learning curve. As the scholars Yip and McKern describe it, China's first big step was a movement "from copying to fit-for-purpose." At the start of the reform period around 1980, Chinese companies (most of them still state owned) essentially copied foreign products such as home appliances, often poorly. Then performance began to rise. Entrepreneurial Chinese firms started to tweak and improve products they had first merely emulated, while contract manufacturers had to meet their foreign clients' requirements, until Chinese goods became "fit-for-purpose" in the sense of being acceptable by local standards.

Gradually, they also began to meet world standards, and then came a movement to the third phase, in which Chinese companies aim for technological leadership. Yip and McKern note that this movement, too, has been driven by multiple influences. They include competition from foreign firms entering the Chinese market themselves, and, as we've mentioned here, the need for new growth as growth from conventional products and industries slows down.

We'll come back to drivers of innovation in the final section on new management models. Now it is time focus on VR/AR.

VR/AR In China: An Emerging Giant?

So far, we've looked at factors that make China a significant player in innovation generally. These same factors are now converging specifically to build the country's growing strengths in virtual and augmented reality.

What stands out most, overall, is that we are seeing top down and widespread commitment to VR/AR in all parts of the vast Chinese economy. VR/AR is literally, and officially, being treated as a big thing in the world's biggest nation.

Here is an initial overview of key forces at work, after which we'll delve more deeply into a couple of them.

The VR/AR push in China has top level backing. President Xi Jinping speaks of its importance, and the central government's current (2016-2020) Five Year Plan includes virtual reality as a focus area for economic growth. In concert with this, a host of government related and government supported initiatives have been launched at levels from the national to the local. Some are radical, such as the striking idea of creating entire "VR towns."

Major Chinese companies are exploring VR/AR. The "Big 3" Internet firms, Alibaba, Baidu, and Tencent, have significant ventures under way. So do mobile phone maker Xiaomi and numerous others. These moves come in addition to existing activity by firms that focus on, say, low cost headset manufacturing or game development. There is also a growing amount of cross border activity and partnering. Most notably, Taiwan based HTC has targeted China as a key ground for sales and development of its HTC Vive products. And to cite another example, California based Jaunt partnered with Shanghai Media Group and others in 2017 to form Jaunt China, a joint venture aimed at, among other things, providing both Chinese and English language VR content for Xiaomi's new headsets.

Not all signs are positive. It has been noted, for instance, that China lags in some areas of cutting edge technology. Many Chinese VR startups have failed, and (as previously mentioned) firms like the Alibaba-Baidu-Tencent trio haven't had much previous success in globalizing their offerings. As one expert told this author, they "haven't found the formula yet for going global."

But the technology is still evolving. Many startups of all types fail. And the big 3 firms have tremendous resources going forward. A news report pointed out that their combined market capitalization is greater than the GDP of Israel. Which leads us to further advantages that China's emerging VR/AR industries can enjoy.

Plenty of money is behind the push. Newer Chinese VR companies received a total of US$543 million in venture capital during 2015-16. Meanwhile the big firms are spending on multiple fronts. Funding their in-house VR programs, creating spinouts that move into VR (like the Baidu spinout iQiyi), and investing in VR startups both at home and abroad.

Total government investment in VR/AR is hard to assess at this writing, but is surely substantial and growing, given the scope of efforts that we'll describe shortly. And last but not least, VR/AR revenues in China are growing rapidly. One forecast projects them to grow to US$8.5 billion by 2020, a hefty number, yet still not the whole story. The revenue projections reflect a powerful underlying force:

Chinese people seem to be falling in love with virtual reality. Already there are more than 3,000 VR arcades spread across the country, many of them placed in malls to draw walk-by traffic from shoppers. These arcades tend to be fully loaded setups, with "stunning booths, that stand out by an extensive use of hardware like special seats and reproductions of vehicles," as a foreign observer noted, while simple stand-alone VR eggs are popping up everywhere, in locations such as bars and coffee shops.

In short, the huge home market that we described earlier in the chapter is translating to a very vibrant new market for virtual reality. Let's start our closer look at these key forces with a bit of detail on this market and what it means for Chinese firms.

Inside China's 'Love Affair' with VR

The VR arcades, in particular, are seen as boosting further growth of the industry in two ways. First, they "democratize" and popularize virtual reality, by making quality VR available to Chinese consumers who can't or won't buy high end gear for home use, but will readily spend a few yuan for half-hour sessions at an arcade. And, the arcades serve as testbeds where content developers and equipment makers alike can find out what appeals to various kinds of customers.

In terms of content, for instance: "Men like a haunted house, and zombie games, while women prefer magic house types. Children like watching the simulation of sceneries," said the young owner of a Shanghai arcade, who reported his

revenues climbing at around 30% per month, with repeat customers coming back and bringing their friends. The same news article on Shanghai's arcade scene revealed other interesting market insights. A mountain climbing game is popular with customers who want to overcome their fear of heights, while the CEO of a content and arcade company said he's aiming to displace pricier forms of entertainment for families: "Not many parents can afford to take their child to the amusement park 10 times per year, but VR can make a child feel like they are at the amusement park."

Moreover, the growing popularity of VR in China could have a long term ripple effect. If, as we saw earlier in the chapter, Chinese people are increasingly tech-savvy and highly educated, and if more young Chinese are now being exposed to virtual reality, wouldn't more of them become likely to go on and work in the VR/AR industries, thus building up the country's talent pool? A young product manager at a VR technology firm in Shenzhen has blogged about being caught up in the wave himself. In a long and thoughtfully researched 2017 post, he wrote of the advances he hopes to see in several areas, as "more and more content creators" come onstream while Chinese companies move from "imitating and catching up with" world class technologies to "improving and creating new ones" to deliver the content effectively.

Now let's turn to what government is doing to support such advances. This will be followed by general commentary and analysis on activity in Chinese firms.

Government Initiatives: Tangible Support for a Virtual World

President Xi Jinping, who was educated in engineering (at Tsinghua University), has emphasized the role that technology must play in China's future. In an oft-quoted speech at the B20 Summit in Hangzhou in 2016, he made it clear that VR will be part of the picture:

Scientific and technological innovation holds the key to development ... The new round of scientific and industrial revolution with Internet at its core is gathering momentum, and new technologies such as artificial intelligence and virtual reality are developing by leaps and bounds. The combination of the virtual economy and the real economy will bring revolutionary changes to our way of work and way of life ...

Adding a personal touch, Xi has visited VR research centers and been photographed tinkering with headsets. In a country where people closely watch the tone set by leaders, that's a subtle but strong message that VR is "cool."

And tangible steps are being taken to drive its growth. The release of the 2016-2020 Five Year Plan has been accompanied by government related VR initiatives all up and down the line. Steps at the national level include creating a true industry ecosystem, with joint public/private sector bodies to identify

Per capita GDP in real US dollars rose from US$348 in 1980 to an estimated US$6,894 in July 2017. During the same period, according to World Bank figures, **the number of Chinese people living in extreme poverty was reduced by about 800 million,** with many rising from low wage status into the middleclass or higher.

and encourage best practices, support and evaluate R&D, and so forth. Two bodies of this type, both formed in 2016, are the China Virtual Reality Industry Alliance and the VR Application Branch of China's Culture and Entertainment Industry Association.

As for steps elsewhere, consider this flurry of activity reported by a Chinese news outlet in early 2016 alone:

Nanchang Municipality has announced it plans to turn Nanchang into the world's first virtual reality city, while Chengdu hosts a virtual reality industry park ... April 1, Changsha held a virtual reality conference, April 9, the Ministry of Industry and Electronic Information Division in Shenzhen City Convention and Exhibition Center organized a "virtual reality industry development forum," and in Beijing the GIC Virtual Reality Summit has just come to an end.

There is much more. Formal central government initiatives include "Internet Plus" and "VR+", with the latter apparently aimed at expanding the use of VR/AR technologies in traditional industries. China's official state owned news agency, Xinhua, launched a VR/AR channel in May 2016. New R&D labs are being formed both by the central government, and by partnerships between private firms and municipalities.

And the most ambitious plans involve building "VR towns," which could host VR/AR companies while also incorporating the latest technologies into nearly every aspect of municipal operation and daily life: Health care, schools, public services, entertainment (of course) and more. One such project is targeted for Donghu, a district of Nanchang City in southeast China. Another project, already in construction, is Beidouwan VR Town in Guian New Area. This one, meant to seed new economic growth for its home province, Ghizhou, is shaping up as a sort of highly enhanced industrial park. It will house companies and workers across a range of VR/AR sectors along with having demonstration areas and facilities for conferences and tourism.

According to an expert interviewed by the author, the government's growing emphasis on VR/AR is producing a double-edged effect. On the one hand, officials at provincial and local levels are moved to propose projects in hopes of securing grant money and scoring "political points" with Beijing. On the other hand, much good is being done. Schools are becoming more VR-oriented; the VR town concept could provide excellent live testbeds for new applications; progress on all fronts is encouraged.

To sum up: Through government involvement at various levels, China is acquiring a very well rounded infrastructure for innovation in virtual and augmented reality.

The Private Sector: Projects and Prospects

VR/AR activities at private firms in China are difficult to summarize, because so much is going on and things change so rapidly. Indeed, the country's IT

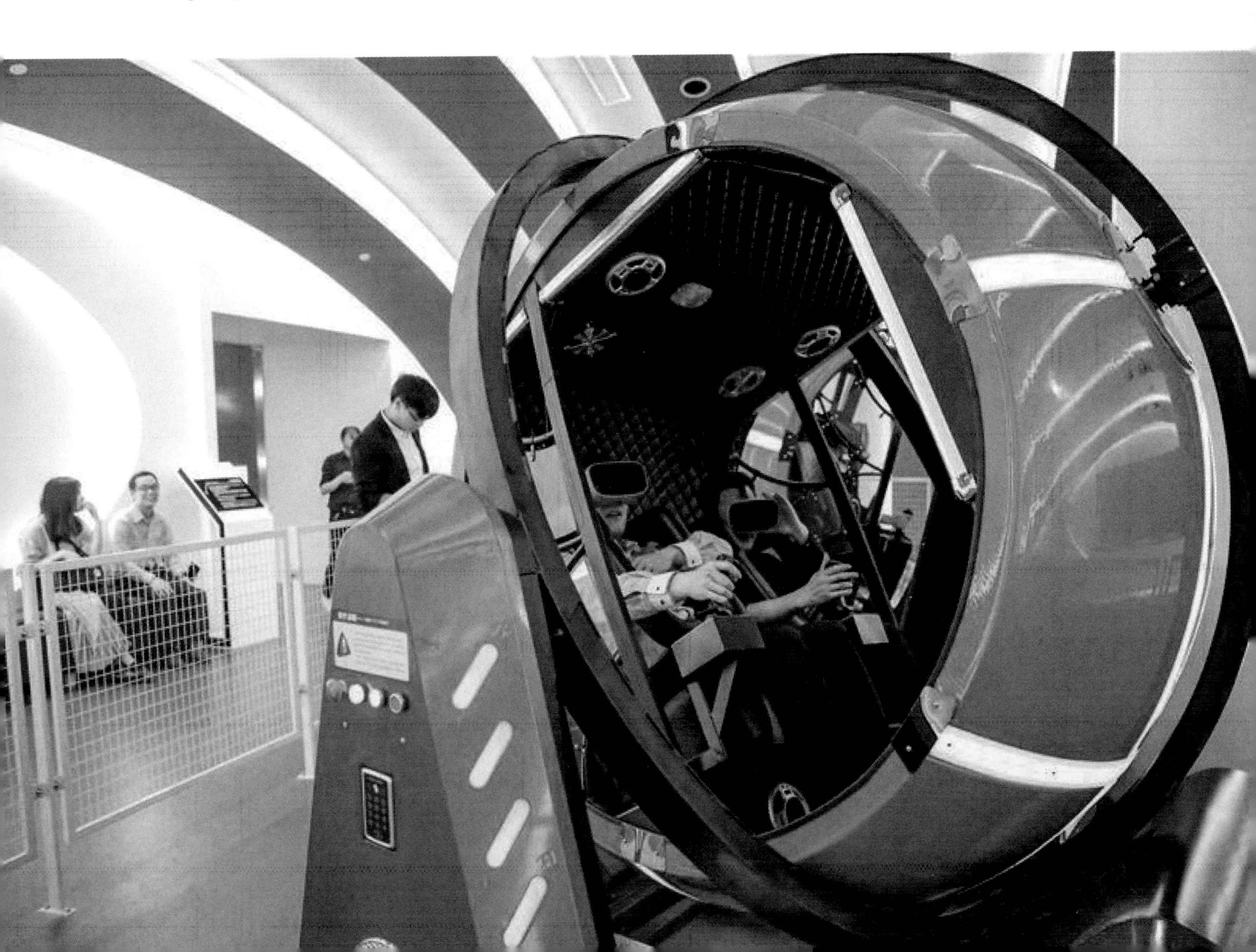

industries are marked by speed. As one source put it, "Chinese companies can offer similar features to their Western counterparts at an accelerated rhythm." So here, we will just try to touch on key aspects of the picture.

To begin with, company initiatives cover a broad range. Alibaba Group, the giant e-commerce operator, is (not surprisingly) developing applications around online shopping. These include giving customers greater ability to examine products through VR/AR, and live streamed VR entertainment in connection with special events like the company's annual Singles Day sale. But Alibaba also has ventures like a newly established GnomeMagic Lab to create VR content for its movie and TV units.

Xiaomi, which grew rapidly in China's mobile phone market by offering high quality, feature rich phones at relatively low prices, now seems to be trying that formula with mobile VR. The company's Mi VR headsets, designed to work with its widely used Mi phones, and some others, have been reviewed as having better than average quality and features, plus a growing selection of open platform content, at a budget price.

There would seem to be a niche for such a product in China. Vast quantities of ultra-cheap wearable viewers are produced there. So many, in fact, that our previously mentioned VR blogger worries they could "decay" or ruin the VR market by giving people a low quality experience. While top of the line equipment remains a financial reach for most consumers. The Mi VR scored strong initial sales by targeting the potential sweet spot in between, but it remains to be seen if such products will prevail eventually.

Altogether, regarding the general near term prospects for VR/AR development in China, expert observers we have interviewed and/or read in secondary sources have put forth the following thoughts:

The hardware sector will probably be a strong point, given the country's experience in physical products plus the emergence of "interesting players" in VR hardware. Companies mentioned in this category included: Lenovo, which among other aspects has partnered with Microsoft and with Disney on content. Also HTC Vive, with its increasing presence on the mainland, plus firms such as Yi Technology (maker of the Yi Halo VR camera, in partnership with Google), headset makers Xiaomi and Deepoon VR, and last but not least, Huawei. The company is a global leader in developing 5G cell networks, which would then enable further advances in other fields, including VR. If Chinese cities are among the first to have wide scale deployment of the networks, as Huawei's 5G chief has predicted they will be, this could benefit China's VR/AR industry generally.

Aside from the booming arcade business, early adoption of consumer VR in China appears to be headed mostly toward mobile rather than in-home systems as in the West. One reason is cost. Another is that the Chinese are keen mobile-phone users, and yet another is that they tend to be status-conscious: it's easier to show off and share your new VR system when you are out and about, whereas fewer people would see it in your home.

Content development is critical. The content, after all, is what users ultimately want, and it's also where a lot of the money will be made. According

VR/AR is literally, and officially, being treated as a big thing in the world's biggest nation. **Chinese people seem to be falling in love with virtual reality.**

to a report in the South China Morning Post, revenue from VR content alone "is forecasted to hit USD$3.6 billion in China by 2021 with over half coming from video and 46% from gaming." A host of Chinese firms are plunging into such content production, both in-house and with partners and spinouts. The entrants range from IT-based companies such as Baidu and Tencent to those with roots in the movie industry, as when China's Sky Limit Entertainment partnered with M2K of France to drive content creation and VR theme park business in both countries.

Content beyond entertainment is being developed in China too. Recall that the VR towns are being planned so as to use and/or demonstrate a wide range of other applications, but how those market sectors will play out remains largely uncertain at present. For now, it seems that immersive video and gaming will be the first content sectors to mature.

Another major unknown, of course, is the degree to which Chinese VR/AR will penetrate global markets. An expert observer offered a very practical take on this subject. With Chinese firms having such a huge home market, "Why would [they] go outside China? ... It is both a lack of incentives and their culture and management style that hinders them from going global."

For the near term, we may in fact see Chinese VR/AR focused primarily on the Chinese market. But we've mentioned various partnerships with U.S. and European firms, and ventures outside China are not limited to those markets. Xiaomi, for one, has been selling its Mi phones quite successfully in India, while Tencent has planted footholds in Indonesia. The Asian and Pacific Rim regions should not be discounted.

Further, it is possible that Chinese companies' "culture and management style" may not be a hindrance to competing globally. Our research is finding evidence that suggests the opposite. Some leading Chinese tech firms

appear to be adopting new management models that could give them a competitive advantage.

A New Management Paradigm Dawning In China?

Increasingly, the world's large companies can be divided into two camps: Those using an outdated form of management, and those that have found a better way. Most firms, by far, are still in the outdated category. They may employ progressive new tools and techniques, but their basic nature is that of an industrial age bureaucracy. They have command and control structures,

elaborate rules and procedures dictating how things should be done, and corporate cultures to match. Growing numbers of business scholars and executives have voiced the need for new models that are fundamentally more dynamic, flexible, and opportunistic. Such models now seem to be emerging. For several years, this author and her colleagues have studied companies that remain entrepreneurial and innovative even as they grow quite large, asking: What are their key management characteristics? And though the research is ongoing, we've identified common threads among selected groups of cutting edge firms.

The 2016 book The Silicon Valley Model described core principles and practices used at Valley firms ranging from Google and Facebook to Tesla. Then our focus shifted to comparative studies of a sample of leading Chinese companies. Initial, early stage research has found them using management approaches that look similar to the Silicon Valley approach in many respects, but different in others and in some ways possibly more advanced. These initial findings are reported in the author's 2017 book Management in the Digital Age: Will China Surpass Silicon Valley?

Breaking New Ground from the Top

For companies anywhere, breaking away from the mold of conventional bureaucracy starts at the top. Visionary, entrepreneurial top leaders are required. In Silicon Valley, these include founder/leaders like Larry Page and Sergey Brin, who famously warned investors in Google's IPO that Google is "not an ordinary company" and would pursue long term innovation over short term maximization of returns. The Chinese companies we've looked at are led by formidable people. They include founder/leaders who have personal track records of visionary innovation, and furthermore have taken steps to broaden and refresh their companies' strategic vision as the firms grow large and diverse. One example is Alibaba's Jack Ma, who recognized the business potential of the Internet before most Chinese did.

Forging New Cultures: Human Factors and a Focus on Speed

What differentiates innovative companies from slower-moving, traditional firms is not just the organizational structure but the culture. The values, norms, and expected behaviors that are established. Facebook's famous mantra is "Move fast and break things." I.e. Don't hesitate to replace successful product features or processes if you've got something better. This is exactly opposite to the old philosophy of "If it ain't broke, don't fix it" and typifies key aspects of the Valley's culture.

Chinese companies have been faced with breaking down mindsets and habits bred by the education system and the state owned enterprises, both of which emphasize deference to authority. The firms have emulated Silicon Valley by setting up open, informal workspaces with onsite amenities, and they've instituted egalitarian practices like insisting that everyone, regardless of rank, address each other by first names or even nicknames. In addition, like Silicon Valley firms, the Chinese companies in our study put effort into recruiting the right people for an ambitious, innovative culture, which means looking for more than good technical skills.

So far, much of this may sound like Silicon Valley translated to China. But one area in which the Chinese firms seriously threaten to outstrip their counterparts is the emphasis placed on speed. In both decision making and product development.

"In the Western world companies talk about 'Fail Fast,' but many Chinese firms have been doing this for years."

Structural Factors: Not as Flat, But Flexible and Open

Silicon Valley firms tend to replace bureaucratic hierarchies and top down management with forms that are more flat, fluid, and distributed. The picture in China seems not quite the same. We find that Chinese firms like those in our study are still more inclined to top down management. Despite the decentralizing and broadening measures we've mentioned, vertical hierarchies tend to persist, and even CEOs, who often are technical experts themselves, may order teams to pursue a development project, then intervene in the details.

Yet there are compensating factors. For example, researchers found that some Chinese firms combine "vertical hierarchy" with "horizontal flexibility" that allows frequent, fast collaboration across units and functions, often in ad-hoc "huddle and act" teams that come together to solve problems or chase opportunities quickly.

Another area where Chinese companies appear to equal or surpass the Silicon Valley firms is the use of open innovation. The Chinese firms avidly pursue strategic partnering in innovation. Baidu, for example, has partnered with both foreign and Chinese automakers on driverless cars, and recently opened its AI-based self driving technology to any companies wishing to use it. A move that one could say echoes Tesla's strategy of opening its patent portfolio.

Conclusions

Chinese companies are shaping up as strong contenders in virtual and augmented reality. They are working in a huge home market that displays an appetite for VR/AR. A number of the firms are being recognized as leaders (or as companies to watch) in hardware development. Many are making investments and forming partnerships in content creation as well. All enjoy the backing of a broad range of government initiatives.

Finally, as shown in our last section, some major Chinese players appear to be using new, innovation oriented management models that could match or surpass the best of breed examples we've found in Silicon Valley, and could put them well ahead of more conventionally managed firms in that regard. Does all of this guarantee that China will "win" in VR/AR? No, but the evidence suggests that in this field, like many others, Chinese companies certainly can't be ignored. ■

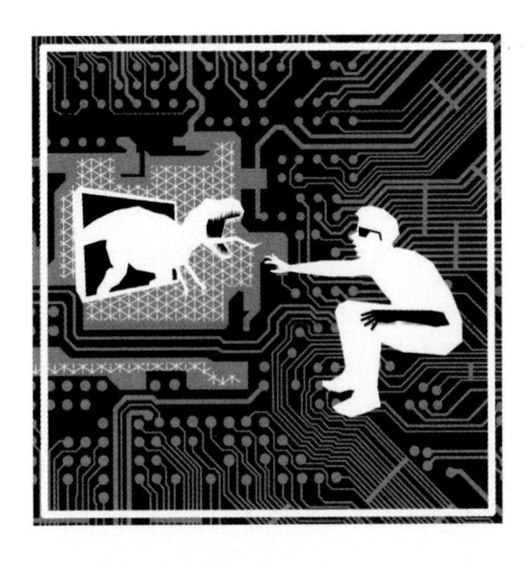

[1] World Intellectual Property Organization (2017): "Record Year for International Patent Applications in 2016; Strong Demand Also for Trademark and Industrial Design Protection," WIPO website, 15 March 2017. http://www.wipo.int/pressroom/en/articles/2017/article_0002.html

[2] Ibid.

[3] Trading Economics (2017): China GDP Per Capita. Trading Economics website, updated 24 July 2017. https://tradingeconomics.com/china/gdp-per-capita/forecast.

[4] World Bank (2017): China home: Overview. World Bank website, 28 March 2017. http://www.worldbank.org/en/country/china/overview.

[5] Eckart, J. (2016) Eight things you need to know about China's economy. World Economic Forum website, June 23, 2016. https://www.weforum.org/agenda/2016/06/8-facts-about-chinas-economy/

[6] Zhang, H. and Sonobe, T. (2011): Development of Science and Technology Parks in China, 1988-2008. Economics: The Open-Access, Open-Assessment E-Journal, 5(2011-6): 1–25. http://dx.doi.org/10.5018/economics-ejournal.ja.2011-6

[7] The Economist (2017): Shenzhen is a hothouse of innovation." Economist.com, 8 April 2017. https://www.economist.com/news/special-report/21720076-copycats-are-out-innovators-are-shenzhen-hothouse-innovation

[8] Yip, G.S. and McKern, B. (2016) China's Next Strategic Advantage: from Imitation to Innovation. The MIT Press, Cambridge MA.

[9] Wu, Y. (2016) Pushing innovation, Chinese firms lead world in R&D spending growth. China Daily, 27 Oct. 2016. http://www.chinadaily.com.cn/business/tech/2016-10/27/content_27185564.htm

[10] Yip and McKern (2016) per note 8 above.

[11] Zhang, Y. and Zhou, Y. (2015), The source of innovation in China: highly innovative systems. Palgrave Macmillan, Basingstoke UK.

[12] Yip and McKern (2016) per note 8, p. 1.

[13] Soo, Z. (2017) Venture capital investments in China surge to record US$31 billion. South China Morning Post, 13 Jan. 2017. http://www.scmp.com/business/china-business/article/2062011/venture-capital-investments-china-surge-record-us31-billion

[14] Ibid.

[15] Chen, L.Y. and Chan, E. (2016) China's local governments are getting into venture capital. Bloomberg.com, 20 Oct. 2016. https://www.bloomberg.com/news/articles/2016-10-20/china-heartland-province-deploying-81-billion-to-seed-startups

[16] See, for example, ibid and Soo (2017) per note 13 above.

[17] Stapleton, K. (2017) China now produces twice as many graduates a year as the US. World Economic Forum, 13 April 2017. https://www.weforum.org/agenda/2017/04/higher-education-in-china-has-boomed-in-the-last-decade

[18] Allison, G. (2017) America second? Yes, and China's lead is only growing. Boston Globe, 22 May 2017. https://www.bostonglobe.com/opinion/2017/05/21/america-second-yes-and-china-lead-only-growing/7G6szOUkTobxmuhgDtLD7M/story.html

[19] Chang, Charles, professor, Fudan University (2017) Skype interview by the author. 13 July 2017.

[20] Strategy& (2017): Competitive China. Strategy& website (undated) at https://www.strategyand.pwc.com/global/home/what-we-think/the_china_strategy/competitive_china

[21] Thompson, C. (2015) How a nation of copycats transformed into a hub for innovation. Wired, Dec. 29, 2015. https://www.wired.com/2015/12/tech-innovation-in-china/

[22] NDTV (2016) Xiaomi says sold 800,000 Redmi Note 2 handsets in 12 hours. NDTV/Gadgets 360, 17 Aug. 2015. http://gadgets.ndtv.com/mobiles/news/xiaomi-says-sold-800000-redmi-note-2-handsets-in-12-hours-728921

[23] Lavin, F. (2016) Singles' Day sales scorecard: a day in China now bigger than a year in Brazil. Forbes.com, Nov. 15, 2016. https://www.forbes.com/sites/franklavin/2016/11/15/singles-day-scorecard-a-day-in-china-now-bigger-than-a-year-in-brazil/#60c500b81076

[24] See for example Clark, D. (2016) Alibaba: the house that Jack Ma built. Ecco, New York.

[25] Chang, C. (2017) per note 19.

[26] See www.baidu.com

[27] Baidu Research (2017) Welcome to SVAIL. Baidu Research website (undated) at http://research.baidu.com/silicon-valley-ai-lab/

[28] Bloomberg News (2017): Baidu Snages 50-Plus Partners for its Apollo Driverless Car. 5 July 2017. https://www.bloomberg.com/news/articles/2017-07-05/baidu-signs-50-plus-partners-for-apollo-driverless-car-project

[29] Yip and McKern (2016) per note 8.

[30] Fischer, B; Lago, U. and Lui, F. (2013). Reinventing giants. Jossey-Bass, San Francisco, p. 46.

[31] Yip and McKern (2016)

[32] Ibid.

[33] Xi Jinping (2016) Keynote speech at the opening ceremony of the B20 Summit in Hangzhou. Transcript from People's Daily, 4 Sept. 2016. http://en.people.cn/n3/2016/0904/c90000-9110023.html

[34] Chang, C. (2017) per note 19.

[35] See for example Robertson, A. (2017) HTC is launching an all-in-one Vive headset just for China. The Verge, 26 July 2017. https://www.theverge.com/2017/7/26/16036100/htc-vive-standalone-qualcomm-vr-headset-china-announce.

[36] Jaunt, Inc. (2017) Jaunt VR App Officially Launches in China on Xiaomi VR Headset. JauntVR.com, 13 June 2017. https://www.jauntvr.com/news/jaunt-vr-app-officially-launches-china-xiaomi-vr-headset

[37] Interview by author, source kept anonymous, July 2017.

[38] Damiani, J. (2017) 90% of Chinese VR Startups Have Gone Bankrupt. Here's Why That's a Good Thing. VRScout, 1 Jan. 2017. https://vrscout.com/news/90-chinese-vr-startups-gone-bankrupt-heres-thats-good-thing/#

[39] Graylin, A. (2017) Skype interview by author of Alvin Graylin, 31 Aug. 2017.

[40] Damiani, J. (2017) per note 38.

[41] Chen, L.Y. (2016) China's Virtual Reality Market Will Be Worth $8.5 Billion and Everyone Wants a Piece. Bloomberg.com, 19 May 2016. https://www.bloomberg.com/news/features/2016-05-15/china-s-virtual-reality-market-will-be-worth-8-5-billion-and-everyone-wants-a-piece

[42] Dayan, Y. (2017) 6 reasons why China is leading virtual reality growth worldwide. Medium.com, 14 February 2017. https://medium.com/@yonidayan/6-reasons-why-china-is-leading-virtual-reality-growth-worldwide -c9a37f4ef2ec

[43] Llamas, S. (2016) How China's biggest companies are crushing it at virtual reality. Develop online, 20 Sept. 2016. http://www.develop-online.net/opinions/how-china-s-biggest-companies-are-crushing-it-at-virtual-reality/0224278

[44] Chen, L.Y. (2016) per note 41.

[45] Dayan, Y. (2017) per note 42.

[46] C-Milk and SerpentZA (2017) China Leads the World in Virtual Reality. Video commentary posted on YouTube, 14 May 2017. https://www.youtube.com/watch?v=Hr4RBHlYrMk

[47] Yoo, E. (2016) Virtual Reality Arcades Are Booming in Shanghai. ChinaFilmInsider.com, 12 Aug. 2016. http://chinafilminsider.com/virtual-reality-arcades-booming-shanghai/

[48] Ibid.

[49] Ibid.

[50] Zhang, Z. (2017) 6 Things You Need to Know About Chinese VR Market. VirtualRealityPop.com, 2 March 2017. https://virtualrealitypop.com/6-things-you-need-to-know-about-chinese-vr-market-ccd8a5c5b85c

[51] Wikipedia (2017) Xi Jinping. Wikipedia.org, edited 4 Oct. 2017. https://en.wikipedia.org/wiki/Xi_Jinping

[52] Xi Jinping (2016) per note 33.

[53] C-Milk and SerpentZA (2017) per note 46.

[54] Liu Shore (undated) Virtual Reality Becomes Prevalent Once More. ChinaGoAbroad.com at http://www.chinagoabroad.com/en/article/20647

[55] See for example Deepoon VR (2017) China's DPVR Launches New PC Helmet E3. PR Newswire, 29 March 2017. http://www.prnewswire.com/news-releases/chinas-dpvr-launches-new-pc-helmet-e3-300431159.html

[56] Liu Shore (undated) per note 54.

[57] Graylin, A. (2017) per note 39.

[58] Dayan, Y. (2017) per note 42.

[59] Ibid.

[60] See for example this (mainly Chinese-language) video: Beijing TVC (2016). Donghu VR Town. Posted on YouTube 21 June 2016. https://www.youtube.com/watch?v=__f8VcbiTXw

[61] Guian New Area (2017). Guian New Area to Establish Beidouwan VR Town to Jumpstart Virtual Reality Opportunities. PR Newswire, 27 Feb. 2017. http://www.prnewswire.com/news-releases/guian-new-area-to-establish-beidouwan-vr-town-to-jumpstart-virtual-reality-opportunities-300413829.html

[62] Graylin, A. (2017) per note 39.

[63] Dayan, Y. (2017) per note 42.

[64] Zhang, Z. (2017) per note 50.

[65] Chen, L.Y. (2016) per note 41.

[66] See for example Charara, S. (2017) Xiaomi Mi VR Review. Wareable.com, 21 March 2017.

https://www.wareable.com/xiaomi/xiaomi-mi-vr-review

[67] Zhang, Z. (2017) per note 50.

[68] Interview sources include Wang, S (2017) Skype interview with Sam Wang of Skylimit Entertainment, 1 Oct 2017, and Graylin, A. (2017) per note 39.

[69] Sin, B. (2017) How Huawei Is Leading 5G Development. Forbes.com, 28 Apr. 2017. https://www.forbes.com/sites/bensin/2017/04/28/what-is-5g-and-whos-leading-the-way-in-development/#4db96c082691

[70] Park, M. (2017) 4 things that make the China VR market unique. UploadVR.com, 5 July 2017. https://uploadvr.com/4-things-make-china-vr-market-unique/

[71] Interview, anon. source per note 62.

[72] Soo, Z. (2017a) Chinese will use 86 million virtual reality headsets within five years. South China Morning Post, 8 June 2017. http://www.scmp.com/tech/social-gadgets/article/2097509/chinese-will-be-using-86m-virtual-reality-headsets-within-5

[73] Chen, L.Y. (2016) per note 41.

[74] Keslassy, E. (2017) Zhang Yimou's Sky Limit Entertainment, France's MK2 Launch VR Venture SoReal. Variety, 24 May 2017. http://variety.com/2017/film/global/zhang-yimous-sky-limit-entertainment-frances-mk2-launch-vr-venture-soreal-exclusive-1202442751/

[75] Soo, Z. (2017a) per note 72.

[76] Graylin, A. (2017) per note 39.

[77] See for example Rutnik, M. (2017) Xiaomi sets sales record due to high demand in India. AndroidAuthority.com, 2 Oct. 2017. http://www.androidauthority.com/xiaomi-sets-sales-record-high-demand-india-804179/

[78] Wu, K. and Zhu, J. (2017) China's Tencent invests in Indonesia's Go-Jek amid SE Asia push: sources. Reuters, 4 July 2017. https://www.reuters.com/article/us-gojek-m-a-tencent/chinas-tencent-invests-in-indonesias-go-jek-amid-se-asia-push-sources-idUSKBN19P17N

[79] See for example Hamel, G. (2009). Moon Shots for Management. Harvard Business Review, Feb. 2009. https://hbr.org/2009/02/moon-shots-for-management

[80] Steiber, A. (2017) Management in the digital age: will China surpass Silicon Valley? Springer International Publishing, Cham, Switzerland.

[81] Brin, S. and Page, L. (2004) 'An Owner's Manual' for Google's Shareholders. Alphabet website at https://abc.xyz/investor/founders-letters/2004/ipo-letter.html

[82] McGregor, J. (2014) Five things to know about Alibaba's leadership. The Washington Post, 18 Sept. 2014. https://www.washingtonpost.com/news/on-leadership/wp/2014/09/18/five-things-to-know-about-alibabas-leadership/?utm_term=.f7333427498d

[83] See for example Kuo, K. (2013) What is the internal culture like at Baidu? Forbes.com's Quora website, 29 March 2013. http://www.forbes.com/sites/quora/2013/03/29/what-is-the-internal-culture-like-at-baidu/#7773a9f72401 and Rabkin, A. (2012). Leaders at Alibaba, Youku, and Baidu are slowly shaking up China's corporate culture. Fast Company, 9 Jan. 2012. http://www.fastcompany.com/1802729/leaders-alibaba-youku-and-baidu-are-slowly-shaking-chinas-corporate-culture

[84] See for example Rabkin (2012) as above and Heng, W. (2014) A peek inside Alibaba's corporate culture. Forbes.com, 13 May 2014. https://www.forbes.com/sites/hengshao/2014/05/13/a-peek-inside-alibabas-corporate-culture/#3862514b4efc

[85] McKern, B. (2017) Skype interview by author, 15 July 2017.

[86] See for example Yip and McKern (2016), and several of the author's interviews have confirmed the observations made here about top-down management.

[87] Williamson, P. J. and Yin, E. (2014) "Accelerated innovation: the new challenge from China." MIT Sloan Management Review, Summer 2014. http://sloanreview.mit.edu/article/accelerated-innovation-the-new-challenge-from-china/

[88] Ibid.

[89] Russell, J. (2017) Baidu is making its self-driving car platform freely available to the automotive industry. TechCrunch, 18 April 2017. https://techcrunch.com/2017/04/18/baidu-project-apollo/)

[90] Musk, E. (2014) All Our Patent Are Belong to You. Tesla website, 12 June 2014. https://www.tesla.com/blog/all-our-patent-are-belong-you

China's VR Themepark

If the enormous selection of various VR experiences and attractions don't get you, the $15 million Transformers statue will.

If there was any doubt that the VR boom in China was all hype, then it's probably starting to fade now. It's been common knowledge for a while now that the Oriental Times Media Corporation has been working tirelessly on a massive VR theme park for Chinese patrons. Well, thanks to the Chinese news media, we finally have a clear look at the still-in-progress complex and the word "massive" is putting it extremely lightly.

Spanning over 320 acres and costing an estimated $1.3 to $1.5 billion, the East Valley of Science and Fantasy will feature multiple attractions, each utilizing VR to help transport users to the future, travel through space, meet extraterrestrial life, ride on the backs of dragons and much more. Users will immediately don their VR gear as soon as they enter the park, including an assortment of interesting peripherals tied to specific experiences. Along with these attractions, the facility will also house a VR restaurant, movie theater, recreational areas, a children's area and China's first VR rollercoaster.

More interestingly, however, are the portions of the park dedicated specifically to VR film production, as well as a media research and development. It appears as though the park, while branded as a VR playground, will actually be doing a whole lot more than meets the eye. It's exciting to see a facility this well funded blend development and entertainment in such a unique way. It's very possible East Valley of Science and Fantasy could become the HQ for all things VR. Not only in China, but the entire Eastern hemisphere. - Kyle Melnick

scan
this

Enterprise GOES FIRST

VR, and in particular AR, have been been building momentum for some time in the enterprise as they enable companies to save tens of millions of dollar in labor in activities as diverse as manufacturing, design, construction, transportation and logistics. While it may be some time before there is critical mass for consumers, there is ROI for business. The inspection time of 60,000 brackets used on an Airbus A380 fuselage, for instance, has now dropped from three weeks to just three days.

The challenges of developing applications for AR are diminished when the technology is being deployed for a single purpose, i.e. inspecting brackets, compared to the multitude of tasks a consumer device would be expected to do. The military needs night vision goggles. They do not need to play music, take pictures, surf the web or keep your calendar.

Finally, the government and the military can afford the newest and best technology years, and even decades, before the enterprise and consumers begin to enjoy their benefits. One thing is clear, AR is making processes the enterprise already does much better, faster and cheaper. It has a measurable ROI and AR is growing exponentially as a result.

By Dirk Schart & Samuel Steinberger

The Workplace OF THE FUTURE

How Augmented Reality Changes The Way We Work and Learn

Dirk Schart is the Head of PR & Marketing at RE'FLEKT, a Munich based tech company building the AR and MR enterprise ecosystem. Dirk is co-author of the book Augmented and Mixed Reality as well as a recognized speaker at events such as Audi MQ, AWE, and SXSW.

Samuel Steinberger is a producer, editor, and journalist whose video and writing bridges the intersections between technology, business, and travel. His work, both visual and written, has been published and broadcast across a variety of outlets, including *Forbes*, *American Banker*, and PBS. He is based in New York City.

VR, and in particular AR, have been been building momentum for some time in the enterprise as they enable companies to save tens of millions of dollar in labor in activities as diverse as manufacturing, design, construction, transportation and logistics. While it may be some time before there is critical mass for consumers, there is ROI for business. The inspection time of 60,000 brackets used on an Airbus A380 fuselage, for instance, has now dropped from three weeks to just three days.

The challenges of developing applications for AR are diminished when the technology is being deployed for a single purpose, i.e. inspecting brackets, compared to the multitude of tasks a consumer device would be expected to do. The military needs night vision goggles. They do not need to play music, take pictures, surf the web or keep your calendar.

Finally, the government and the military can afford the newest and best technology years, and even decades, before the enterprise and consumers begin to enjoy their benefits. One thing is clear, AR is making processes the enterprise already does much better, faster and cheaper. It has a measurable ROI and AR is growing exponentially as a result.

October 2028: It's a typical Monday morning. As soon as you get out of bed, information starts streaming in. Glance out the window: Hovering on the New York City skyline are holograms informing you about the weather, last night's game, or how many minutes of cardio you'll need at the gym later today. While you're enjoying a cup of coffee, floating projections show the morning news, your first meetings and the latest messages. Your virtual fitting room presents several different outfits for the day, letting you choose the perfect look. Then it's time for your morning meeting, which starts halfway into your commute. The autonomous car that picked you up, now navigating the streets of Manhattan, notifies you of an incoming conference call. It's your team: One colleague in Berlin, the other in Hong Kong, and with a quick swipe, both avatars are virtually present in the back seat of your car.

Anyone who has seen more than a few science fiction movies can attest to the frequency in which films begin with scenes like this, setting the tone for a world far more advanced than our own. And while many might believe this world only exists in some far off future, we are getting closer everyday. Science fiction is becoming science fact

Augmented Becomes Reality

In the past five years, augmented reality (AR) developed from a technology used by researchers, scientists and a handful of nerds to a useful enterprise tool with the potential to change the way we work and learn. Apple, Facebook, Google and Microsoft have invested heavily in AR, building tools like ARKit and ARCore, providing developers the infrastructure needed to build AR

experiences for iOS and Android devices. On the hardware development side, Microsoft's HoloLens glasses and the enterprise edition of Google Glass, among other AR glasses, are robust in their design. The technology giants are paving the way for this technology to reach millions of users and developers.

Now that AR has some time to develop and gain traction, enterprise users are moving beyond first pass tests and novelty apps to new business models that leverage the hardware improvements found in the latest generation of smart glasses. To be clear, we're still in the early stages. But the technology industry knows that the PC is disappearing. We are currently in the mobile age, and ultimately this is a stepping stone toward ambient spatial computing.

When AR converges with artificial intelligence, and a more intuitive user interface emerges, allowing us to interact with data in a natural way, we will have arrived at the workplace of the future. We're close, but AR glasses still only provide useful assistance for specific tasks. When we reach the point where we can multitask in the augmented workplace, we will be able to collaborate better and solve several of the problems we face all day.

Perhaps you're looking for a more visual way to work. Anyone who's tried to manage interdisciplinary teams and partners in different countries knows that most meetings are global affairs with online calls and conferences. If you want to share the latest draft, you need to take a laptop, connect to a big screen for the benefit of the people in the room, and transfer your screen to an online interface to show your global colleagues what you're seeing in your office. And it's still all presented on a 2D screen.

Furthermore, interaction is limited by switching between content and colleagues. Why not just put on your AR glasses, connect the team and share the content? All participants will see the same content, and some might even decide to walk around it to get a different or better angle. Imagine you want to plan a new store. Present a new car design. Or guide your management team through mountains of connected numbers and analytic data. You can handle virtual objects and drag them around to illustrate ideas.

For those who travel frequently, and work on the go, one little screen is uncomfortable, as it is inefficient. With AR glasses, the whole concept of browser tabs becomes obsolete. Screens can be placed in the physical world, and moved around in physical space. The Microsoft HoloLens headset allows users to bring their windows desktop onto the device with full access to the office suite of Powerpoint, Excel, and Word. A virtual desktop can travel with us anywhere.

Unfortunately, today's smart glasses can only do what you tell them. They are not the intelligent assistants we want them to be. The user still needs to teach the glasses about the environment. The geospatial AI that can instantly

When we no longer require a physical keyboard and 2D screen is when **we've achieved the holy grail of the user interface**

detect the geographical locale and recognize objects, people and places, does not exist yet. Once a device knows where it is, it can place the content in the right place, at the right time.

The Journey to The Workplace of the Future

After your morning commute and meeting, you arrive at the office. On the schedule is a workshop to discuss new designs. While walking to the room, your virtual assistant shows you the current drafts, your notes, and some background on the meetings' participants. You stop to grab another cup of joe, but the coffee machine isn't working. You're confused. You're no technician, and you don't have the time to read the coffee machine instructions. But in 2028, there's no need to read instructions. Your assistant has already registered the coffee machine model and the error via the connected sensors in the machine. Visual step-by-step instructions appear on the machine guiding you to change a filter. Your virtual assistant knows which drawer the replacement filter is in, and what you have to do next. Success! Within five minutes, you have coffee in hand, instead of having to call help you get your caffeine fix.

Tomorrow's AR glasses will be even more functional than today's digital assistants, because they'll recognize visual components in the world around you, providing context for content. Today's smart glasses are elementary. Devices like Google Glass, Epson Moverio or Vuzix, are designed to provide 2D content. Although their usefulness is limited, their batteries often last longer and they're lighter and better to wear for a longer period of time. Further along the spectrum, there is the Microsoft HoloLens, ODG's R-8 and R-9, as well as the Meta 2 (not to mention Magic Leap will soon introduce their own AR glasses). These devices are bigger and more powerful for 3D real time visualization. For the workplace of the future, we need smart glasses with even better form factors, longer lasting batteries and at much lower prices. In 2028, I expect stylish glasses, as comfortable as sunglasses, and for a price less than $1,000.

The HMD that will dominate has not yet emerged. There are several approaches to problems like miniaturization being tested. Some AR HMDs,

Courtesy Case Western Reserve University

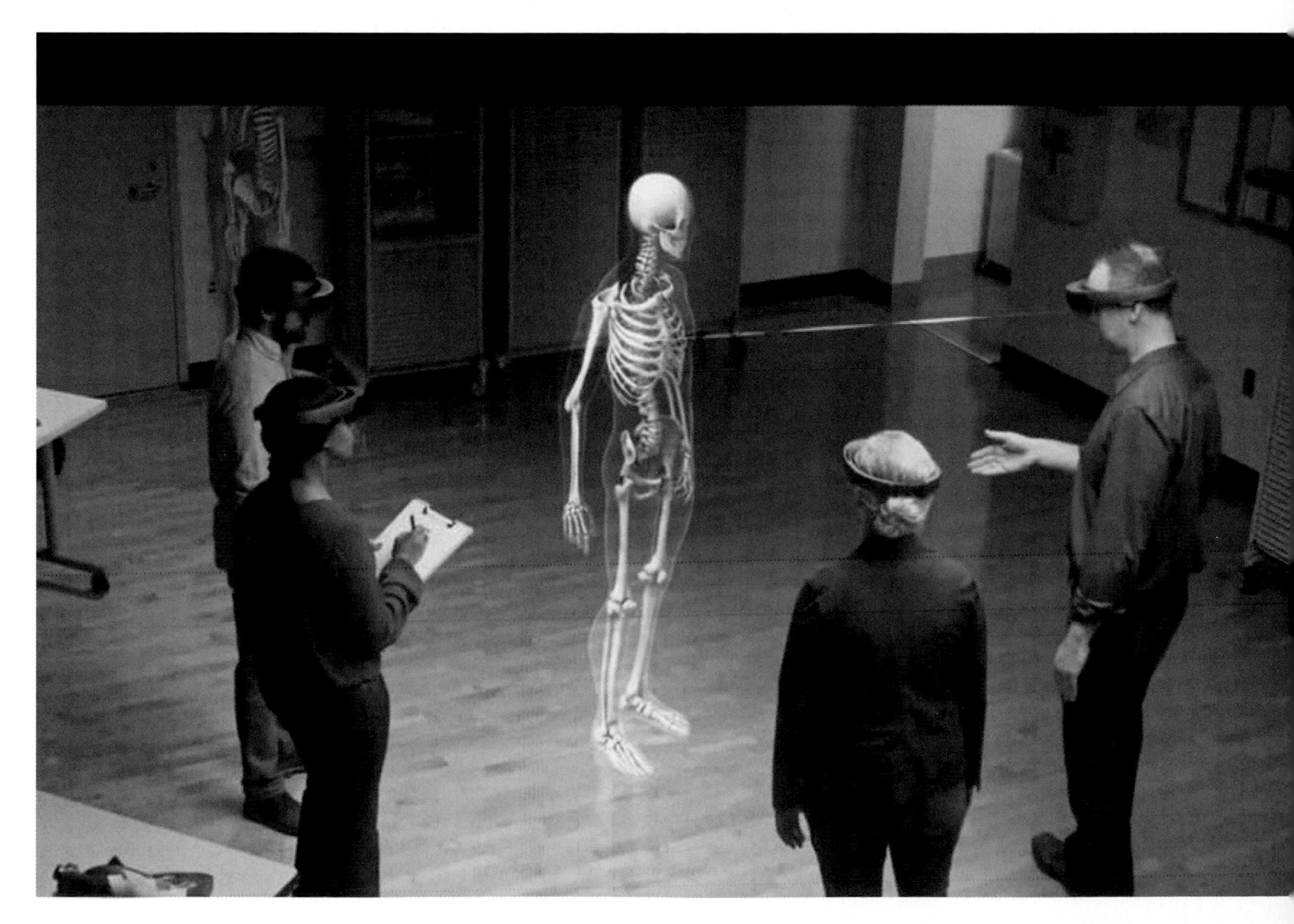

like the Ousterhout Design Group's (ODG) R-8, are bulker than sunglasses, but they have all the hardware on board. Other HMDs, like the Meta, are tethered to a PC, which restricts mobility. And yet other devices will have a little belt pack for the battery and processor, which is how it is said Magic Leap and Apple HMDs might work. Another option is to have everything built-in for simpler tasks without a lot of performance. Different tools are needed for different tasks. Therefore, there will be a wide variety of AR HMDs at varying price points and options, as there are with smartphones.

Besides the hardware itself, I also see other important areas for the perfect AR experience: A realistic visualization of virtual objects, natural user interfaces, intelligent recognition, and tracking of the physical environment and real objects. The key to a realistic visualization that merges the digital and the real world is how the content is displayed.

Companies like Magic Leap or Avegant are working on solutions with light field technology where the content will be projected directly into your eyes. First prototypes show that they're able to simulate how our brain "sees". If you focus on an object, the background or other objects get blurry. If you focus on the background, the same happens with the object in front of you.

What Will The Workplace of the Future Look Like?

In 2028, you'll be wearing your glasses as you walk around buildings or supermarkets, and since companies will know exactly where you are, they'll be able to show you information as you interact with the real environment. A light vibration on your wrist and an error warning in front of your eyes alerts you to a broken machine. Blue arrows on the floor guide you to the culprit. As you get closer, you're automatically logged in and a personalized profile loads. Holographic elements form around you: Machine status, sensor data and a troubleshooting guide. Within seconds you know what to do. Your eyes go to the left, triggering the information overview to minimize. With your hand, you drag the repair imagery into place, overlaid onto the machine.

In 2028, workers will be supported by intelligent assistants and smart glasses, working hands free. Instead of wasting time searching for the right information, all instructions and data will be in your field of view based on where you are and what you're looking at without the user doing anything. The key interface here is hands free. This is why there is so much investment and development in products like Amazon's Alexa and Google Home. When we no longer require a physical keyboard and 2D screen is when we've achieved the holy grail of the user interface.

Today, we're in a transition from mobile devices to smart glasses. AR has started with smartphones and tablets, as they are readily available and as a result, many use cases have already been built for them:

Collaboration and Support

Hyperloop Transportation Technologies (HTT) is a Los Angeles based transportation company building the Hyperloop, a magnetic levitation system (envisioned by Elon Musk) in a vacuum tube running at the speed of sound. The startup, founded in 2013, works with internationally distributed teams which presents a myriad of collaboration challenges that are solved, in part, with VR.

Running the first version of the Beam app on smart glasses, HTT has a virtual office with access to different projects and documents, streamlining communication so that team members can explore and discuss 3D content together.

the technology industry knows that **the PC is disappearing**

No longer restricted to their offices, team members with smart glasses can work in the virtual Beam workshop from just about anywhere.

Training and Education

Training today is carried out as it has been for centuries. Workers learn on the job as an apprentice, and are tutored by colleagues and in more formalized settings, slowly gaining proficiency. However, this on-boarding process has a cost in time and efficiency. Often, the trouble is that the equipment is in use, as it is for Siemen's client Deutsche Bahn, the German national express railway, which does not have extra rolling stock for training. Siemens created a VR solution that the company now touts as an example of the efficiency and cost savings that VR now offers the enterprise market.

Medical device manufacturer Getinge makes complex healthcare devices. Doctors and clinical personnel need to be trained in its configuration and setup. To solve this, the company developed an AR program paired with an HMD, allowing doctors and nurses to train with their own equipment in their own hospitals without traveling to a dedicated seminar. Trainees from different locations participate in the training at the same time and repeat the course whenever they want. Remote experts are also available.

Car manufacturer Range Rover wanted to show trainees the sensors and wires behind the dashboard of their vehicles, however, this meant that the entire dashboard needed to be dismantled. This tended to cause damage to an expensive vehicle each time a trainee opened the dashboard to learn about the cable infrastructure.

ENGINE REPAIR WALKTHROUGH
Remove the top cover
Track down the error area
Place the bolt into the bolt hole at the marked location
Use an open-end wrench to tighten up the bolt.
repeat the process on the next marker

1
Noise Level: 82dB
Vacuuming Performance: 2871
Temperature: 42.1
Motor Vent Inspection: Done
Leak Inspection: Done
STATUS: OK
2
Noise Level: 130dB
Error Code:N22673
SYSTEM ERROR!
3
STATUS: OK

How can you visualize what's hidden? You sit in the car, put on your HMD on and check the dashboard or the doors. You can look behind the virtual surfaces to see the wire configurations and sensors. Is that sensor flashing red? There might be a problem with it, and with one gesture you get more information about the problem, and better understand the steps needed to fix it.

Visualization and Simulation

Every day we collect mountains of data and put it somewhere in the cloud. Then, when needed, we waste time just finding and interpreting the data. AR can visualize data directly in the worker's field of view, overlaid directly onto relevant areas, such as air pressure levels shown directly on a tire. Industry giant Bosch recently demonstrated how they'd like to bring the car garage into the digital age. One approach is to support technicians with real-time data coming from car sensors, converted into an AR-friendly projection for the repair work to take place. Taking it one step further, that same information can be combined with remote services, supporting car owners who have broken down on the highway or are having car trouble at home.

In Bosch's vision, you walk around the car with your HMD. The computer and the sensors in the car show their status and visualize a possible malfunction. The car mechanic can review it together with you, automatically send the right information for the needed parts, and provide repair instructions.

Instruction and Guidance

As our machines get more complex, access to specialists is essential. Having experts on call is cheaper than having experts on staff. Remote experts see what the worker sees in real time, and can instruct and even draw in the worker's field of view, leaving the worker's arms free in the meantime to physically have his hands on the problem. As factories are automated and run increasingly sophisticated equipment, often in far flung places around the globe, this takes on much greater importance.

Virtual assistants can guide workers step-by-step through complex tasks to avoid mistakes. Vacuum pump manufacturer, Leybold, is now using this approach to help in the service and repair of their vast array of pumps, each with different operations and repair manuals. With vacuum pumps on location approach to help in the service and repair of their vast array of pumps, each with different operations and repair manuals. With vacuum pumps on location around the globe for a variety of customers, Leybold wanted to encourage their customers to maintain them on their own.

Recently, Leybold service technicians started to test AR support with tablets and HMDs. Since then, they have developed apps for different pumps in operation and are now preparing to provide all their customers with AR as an additional service. When a customer opens the app, the correct documentation launches on the device. There's the option to choose "X-ray vision" to see all the different parts inside the pump, get detailed information or order parts as needed.

Augmented Reality Creates the Jobs of Tomorrow

Every industrial age to date has changed our working world and our jobs significantly. From skilled craftsmen and to mass production. From the first assembly lines to the computer age: Humans have adapted to new work environments and learned new skills. The next change will be characterized by high amounts of automation, machine learning and artificial intelligence. Many jobs will disappear due to increased automation and robotics. ■

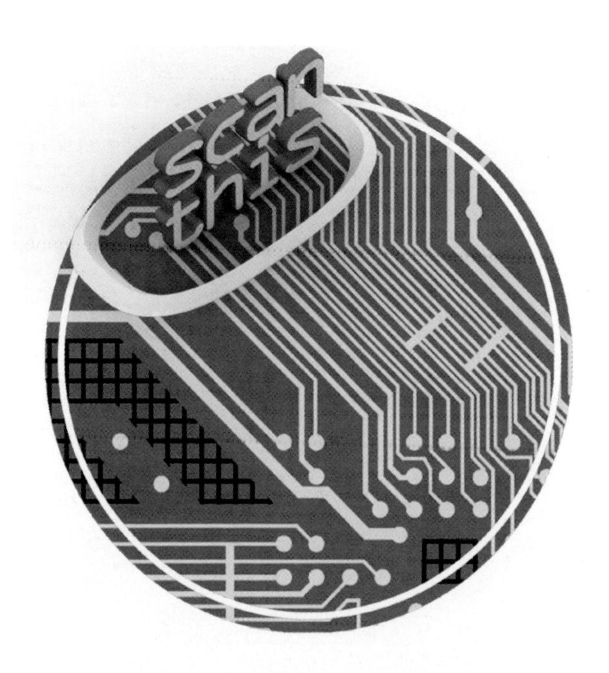

By the time I spotted the shotgun, it was too late. I feel a sting in my thigh. I've been hit. I was distracted by the hysterical wife, while my partner was dealing with a nosy neighbor. We get off thirteen shots. All high. Merely grazing the assailant. The simulation comes to a stop. "Rookie mistakes," says our trainer, Deputy Jose Diaz of the LA County Sheriff's Department (LASD). The LASD is a 17,000 man force that polices the endless suburban sprawl outside LA's city limits.

The shotgun was on the shelf behind the subject the whole time. We missed it. At the controls of the simulator, made by VirTra, James Grady, who operated the branching 360 narratives in real time, smiles. It could have gone another way. The subject could have complied. "What fun would that be?" He smiles. The simulator is for advanced officer training. "This is much more realistic than range training." Scripted live simulations with actors are the traditional way to accomplish this, but it's difficult and expensive. Debriefings are limited to after action analysis. With VirTra, the action can be stopped and replayed. Grady explained that "during our live training, deputies' heart rates go through the roof. This isn't as intense, but we can still get you going pretty good in here."

The VirTra system consists of a 300 degree, five sided screen, like a wrap around Cave Automatic Virtual Environment (CAVE). Trackers follow the participants. The floor is wired for sound and vibration. For security training, VR without a head mounted display is much more realistic, because that's the way it is in real life. Real weapons, modified with lasers, and supplemental equipment like tasers and mace can also be deployed within the simulation. The deputy is equipped with the exact same equipment they have on patrol, allowing them to make use of force choices they'll have to explain later. The system plays branching live action 360 videos, which an operator chooses based on the responses of those in the simulation. VirTra provides most of the content, but the LASD also worked with the company to create custom content. "Those entrusted with lethal force decisions should be provided training equal to the importance of the decisions they must make," said VirTra Founder and CEO Bob Ferris.

Public Safety:
A Case Study

Dr. Walter Greenleaf

VR and Health Care

Walter Greenleaf, PhD is a research neuroscientist and medical product developer working at Stanford University. He is known internationally as an early pioneer in digital medicine and virtual environment technology.

Virtual Reality Technology for Medicine

Virtual reality technology has been used for more than two decades to support medical training, improve clinical interventions, and to promote health and wellness. VR clinical systems have been designed, developed and validated primarily at university research centers, and as a result, there is an impressive body of published research literature demonstrating the principal efficacy and therapeutic value of virtual reality (VR) based tools in medicine, providing the proper context for further advances.

Until recently, the cost of VR technology has been a barrier to acceptance outside of the university environment. Now, however, the reduced cost and increased availability of virtual reality technology have made clinical VR practical for everyday use. As a result, VR technology has migrated from the university and surgical training centers to other sectors of medicine.
Leveraging Training and Gaming Paradigms

VR technology made its first inroads into widespread use through the computer game industry and the military, where it was used in flight simulation and other training scenarios. Applications in medicine shortly followed, with VR seeing increasing use as a medical teaching tool, for planning and conducting surgical procedures (including remote surgery), and for interactive diagnostic imaging. In the realm of behavioral medicine, VR has now seen several years of successful use in the treatment of PTSD and phobias.

Recent technical advances in the graphical power of VR systems have provided dramatically increased levels of immersion and "presence" in VR environments, opening up new possibilities for the use of VR in behavioral medicine. In the hands of trained clinicians, VR tools offer great potential for increasing treatment efficacy and reducing treatment time. Moreover, the use of VR in a telemedicine context enables the expansion of clinical reach to underserved populations.

A Broad Impact on Healthcare

All aspects of health and wellness are being impacted by the application of this next wave of technology. In addition to clinical skill training, and as a way to evaluate patients in a more natural/objective manner; virtual reality technology is being used in the fields of psychiatry, clinical psychology, physical medicine and rehabilitation.

Six Major Areas of Impact

- Prevention: Promoting Health / Wellness Habits
- Improved Training
- Objective and Accurate Assessments
- Improved Interventions
- Improved Adherence
- Distributed Care Delivery / Telemedicine

portable VR technology can be deployed **as the foundation for a telemedicine** program.

Case Study: Treating PTSD with VR

Skip Rizzo (above) associate director for medical virtual reality at the USC Institute for Creative Technologies, has been working with the U.S. Army on ways to use virtual reality (VR) to treat soldiers' Post Traumatic Stress Disor-

Photo Courtesy Applied VR

der (PTSD) for over a decade. His system, Bravemind, initially funded by the Department of Defense in 2005, can accurately recreate an inciting incident in a war zone, like Iraq, to activate "extinction learning" which can deactivate a deep seated "flight or fight response," relieving fear and anxiety. "This is a hard treatment for a hard problem in a safe setting," Rizzo told me. Together with talk therapy, the treatment can measurably relieve the PTSD symptoms. The Army has found "Bravemind" can also help treat other traumas like sexual assault.

Bravemind is sophisticated enough to insert minute details of a PSTD patient's inciting incident. How many people are in the vehicle? Where are they sitting? What kind of road are they on? What time of day is it? Or are they on foot in a bazaar? Is it crowded with civilians? Are there vehicles present? Civilians? Helicopters overhead? All these details can be added to the simulation. The operator can play the scene forward and back, slow it down, or freeze the frame, all reflecting the patient's narration. The Army is currently spending two million dollars on a seventy site clinical research trial of Bravemind.

"The VR helped me," said Steve Healy, who served in Iraq, "I remember the simulation better than the original incident at this point. It's taken some of the edge off. I feel like I have fewer, shorter, less intense episodes." The remarkable results were recently published by the The Journal of Anxiety Disorders. Rizzo received the American Psychological Association's 2010 Award for Outstanding Contributions to the Treatment of Trauma.

Once a lone, visionary figure in the world of psychology, Rizzo has fathered an industry. His foresight in understanding how new VR technology might be used to treat mental illnesses through simulation is now being emulated by dozens of companies trying to operationalize the use of VR therapy and training.

Case Study: Pain Management

Josh Sackman, President of AppliedVR, says he was inspired by Rizzo's pioneering work. AppliedVR uses the mobile Samsung Gear VR to treat hospital patients suffering from anxiety, phantom, acute and chronic pain, memory loss, as well as the tedium of a long hospital stay. The content of the VR elements can range from games and guided relaxation to nature experiences and education. Sackman says games are most popular with kids. "Bear Blast" reduces pain in young patients by 24%, a recent trial revealed. "Pain distraction is often hard to achieve in a hospital," said Sackman, who says VR can fill an important unmet need in the treatment of patients. "VR is completely immersive and transportive, an extraordinary experience for patients with little or no mobility."

AppliedVR just raised a three million dollar seed round, which will enable them to train more hospitals in the use of its system. The company also develops

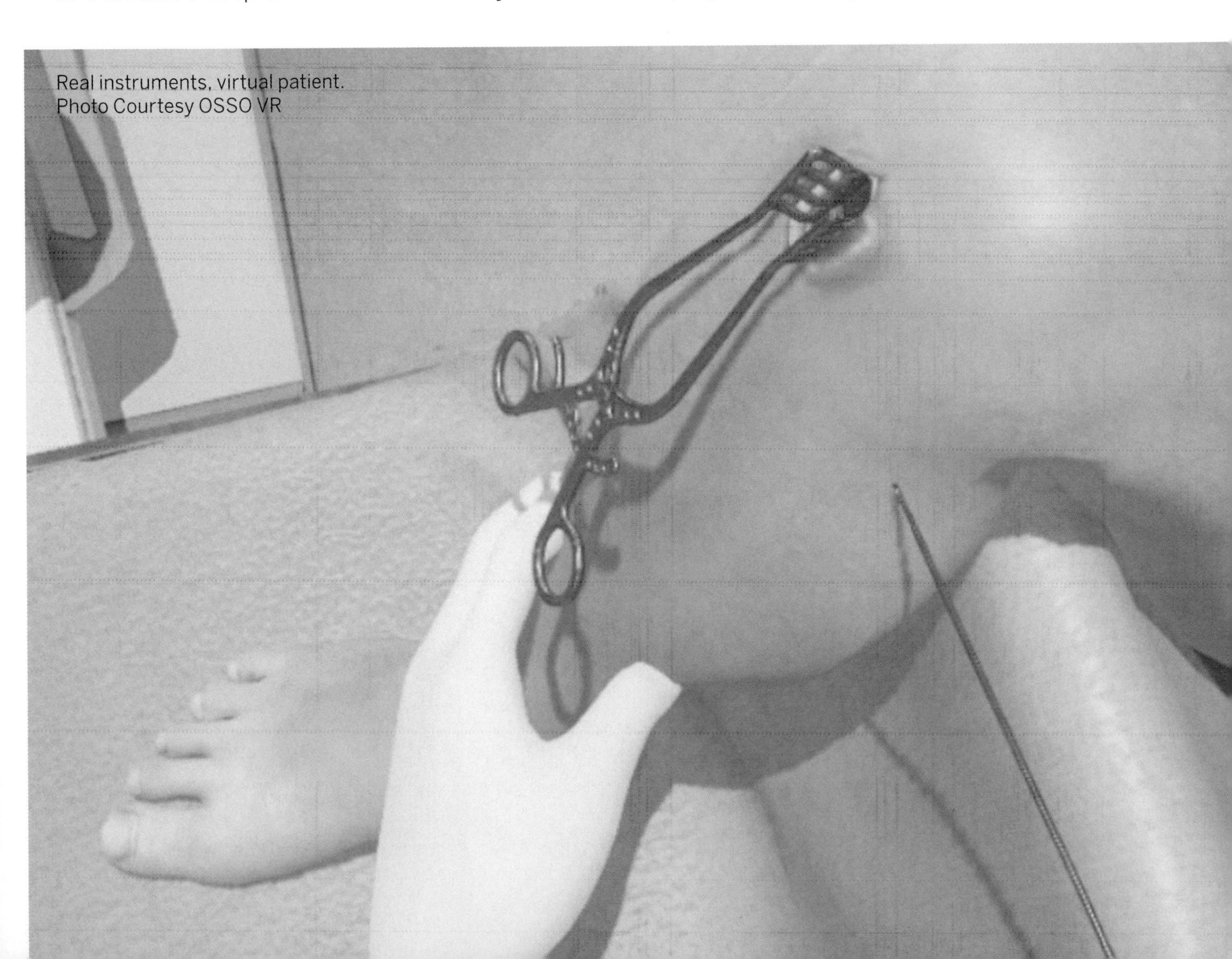
Real instruments, virtual patient.
Photo Courtesy OSSO VR

custom content like "Bear Blast" and curates content from partners. But Sackman says third party content providers will always be the most important part of their offering. "This is like Netflix," said Sackman, "a little of ours, a lot of everyone else's. It really takes a village to do something like this."

AppliedVR has had to overcome a number of obstacles to be accepted in a hospital setting. "It has to work out of the box," said Sackman. "It can't be more complex than launching an app preloaded onto a phone." In addition, the headsets must be sterile. As part of the "kit" provided to the hospital, AppliedVR includes a healthy supply of disposable foam face pads, which can easily be replaced to ensure the Gear VR is as close to sterile as possible outside an operating room.

An immersive virtual reality experience can commandeer a patient's brain so it no longer focuses on pain, says Dr. Brennan Spiegel, a Cedars-Sinai researcher, gastroenterologist, and professor of medicine. "It doesn't work on everybody, but when it works, it really, really works," he said.

Case Study: Surgical Training

Justin Barad, the CEO and Founder of Osso VR, is a practicing pediatric orthopaedist. In a previous life, he was a game developer. Barad knows first hand the challenges orthopedic surgeons face, especially when learning new techniques and devices, and he's drawn on his background in interactive media to create important new tools to train surgeons. "We surgeons operate with a fallacy of transferability, which means we come to believe that because we're good at one kind of surgery, we will be good at another. The data says we are starting fresh, just like a resident, no matter what our experience with other tools and approaches," said Dr. Barad. This is one of the reasons it's so hard to introduce new techniques and devices in clinical medicine. It's too hard and, in some ways dangerous, for a specialist to learn it by doing it.

Hospital workers, long known for graveyard humor, often joke about accident victims and the homeless who require emergency surgeries are "MRBs", short for Medical Resident Benefits. Residents are highly educated and observe many surgeries, but they only operate on one cadaver before doing surgery on a patient. Their results are 300% worse than experienced surgeons. "In med school, they say 'see one, do one, teach one'. It's a confidence builder, but the truth is you need to do fifty to one hundred cases for proficiency," said Barad.

Osso VR uses virtual reality to enable surgeons to perform realistic orthopedic surgery, replacing an old, complicated, expensive manual simulation created by the aptly named Sawbones Corp. Sawbones provides a realistic molded plastic model of the bone for orthopedic training and practice. The models are expensive and can only be used once. Much of Osso VR's business comes from medical device manufacturers, who now have a way to show surgeons the benefit of new implants and techniques while training them in their use. Studies show surgeons who train with Osso VR's system achieve test results twice as good as those trained using current methods.

Distributed Care Delivery / Telemedicine

By leveraging the fact that a computer generated environment enables VR facilitated assessments and treatments, portable VR technology can be deployed as the foundation for a telemedicine program. Cellphone based or standalone VR systems can be used to extend the reach of the clinician, provide healthcare access to underserved populations, support managed home recovery, enhance chronic disease management and protocol adherence. The possibility for the clinician and the client to physically remain in their respective locations while meeting in the virtual space for therapeutic interventions can transform the current focus of clinic based care by shifting the center of care to the patient's location, no matter where that may be. Moreover, the portable nature of VR technology provides the opportunity for therapeutic interventions to be provided early. The emphasis can be placed on prevention or early intervention before disorders become severe and chronic.

The Next Generation of Medical VR

A new generation of medical VR technology is evolving, incorporating multi-user social virtual worlds, aided by natural user interfaces that can detect facial expressions, hand gestures, and body movements. These next generation diagnostic and treatment systems are beginning to make the transition from the research laboratory to general clinical care, providing improved patient outcomes and reducing health care costs. ■

scan
this

LIFE WITH VR & AR

We've talked about a wide range of topics, from man's twin desires for augmentation and immersion, to China's emergence as a technology superpower. Now let's explore how these immersive technologies are already slowly seeping into every aspect of our lives. This has barely begun but, as with Killer Apps discussed earlier, there are clues to how this technology will enhance the things we do every day.

VR and AR's KILLER APPS Will Be Social

By Charlie Fink

The success of social media is an affirmation of our need for human connection and belonging. This need will follow wherever our technology takes us. If people spend a lot of time on Facebook using their phone, they will want to spend a lot of time in a future social environment, whatever that means, Social VR, or maybe AR.

What is social VR?

So what is social VR? What will we do there? Meet strangers (artfully called "social discovery") as you might in a massive multiplayer online role playing game (MMORPG) like *World of Warcraft (WoW)*? Or playing Xbox one on one? What will it cost? If it's free, why is it free? Steve Ballmer famously said "If you're not paying for a service, you're not the customer. You're the product." Who is making money? Even free things have a cost: Our attention.

The holy grail for Facebook (owner of Oculus) is not games, but something much, much bigger: A three dimensional virtual Facebook home page, essentially your virtual home in cyberspace. Social media is something everyone is already doing. But in VR, no one really knows how this will work exactly. Early adopters are leaving clues in places like Second Life, AltSpaceVR, Sansar, High Fidelity and Rec Room.

Games Are the Oldest Form of Online Social Media

Massive multiplayer online games (MMOGs) date back to the formative days of the Internet. The most popular of these, World of Warcraft, a MMORPG from Blizzard Entertainment that launched in 2004, has generated over two billion dollars in revenue. More than one hundred million accounts have been created over the game's lifetime. At its peak in 2010, WoW, as it's known to fans, had over twelve million paying subscribers. Players can quest on their own, or form a team, elevating the social aspect of the challenges. Players like the "social discovery" team play engenders, and bond over the games. Experienced players will tell you they have met real friends in the virtual world. A massive economy developed inside the game, with advanced players selling tokens they earned to those desiring an easier, faster way to acquire special powers. According to Superdata Research, MMOGs continue to be a huge category, representing 27% of the $96.5 billion game market.

Second Life is Social Media

Second Life, the world's best known social virtual world, was founded by Philip Rosedale, the former CTO of Real Networks. He started with Linden World, which was expanded and renamed Second Life in 2003. It was different from the other MMOGs, and it received a lot of attention as a result. It was unstructured. There was no goal. No competition. Users can build their own 3D domains and control virtual real estate. Linden Labs, the game's creator, encouraged the development of an economy with Linden dollars, a currency with an offline equivalent. Linden dollars is now the basis for an annual economy valued at $700 million annually. Second Life made the cover of Business Week in 2006, but never achieved the financial success of a WoW.

It would be a cliche to characterize the players or residents, as they are called, as anti-social nerds who hide behind the anonymity of an avatar. 800,000 people (called residents) log on monthly and take their "second life" very seriously. "People are doing everything you can imagine in Second Life," Linden Lab's CEO Ebbe Altberg told me. "Everything you are doing in real life, people are doing in Second Life." The University of Texas has a chemistry lab there. A politician had a campaign office. Some even make a living there, which they earn through long hours of graphic design and relationship building. I met a Greek girl who said she makes money by singing in Second Life. Arena Stage in Washington, D.C., a regional theater with a national profile, has the VR version of its iconic building by Bing Thom in Second Life. The Jewel Theater performs dramas made especially for Second Life. Working within the confines of the medium, they use text instead of voice (I'll explain why below). The residents I spoke with as I researched this piece were kind and forthcoming. Enough so that I am now thinking Second Life is more of a social network that happens to be in a complex 3D virtual world.

Telling A Story in VR

Penrose Studio's VR experience Allumette was the talk of the Tribeca Film Festival in 2016. Wired Magazine declared "The stunning Allumette is the first VR Film masterpiece!" This fifteen minute narrative, directed by Penrose founder Eugene Chung, is loosely based on a Hans Christian Andersen fairy tale, "The Little Match Girl". Set in a floating Venetian city nestled in the clouds, people navigate among buildings with gravity-defying boats. Allumette has the feel of a popup children's book, which makes it all the more shocking when Chung soulfully and unflinchingly executes Andersen's tragic ending, which we witness painful steps away.

In the early days of film, or motion pictures, or movies, people talked about the new medium as moving pictures, and picture shows. It took some years for people to agree on calling the new medium "movies". We are so early in the development of VR, we don't have words to describe Allumette. It's not a film. It's cinematic, but not live action. It's a story. A narrative. It has main characters, though they are stylized like the stop motion maquettes featured in The Nightmare Before Christmas and James and the Giant Peach. Certainly, like all VR, we experience it. I suppose this is why I find myself and other writers referring to VR experiences. We don't have a better word for Allumette.

Allumette gives the viewer control over scale, perspective and presence, replacing the traditional cinematic language of intercutting and parallel action invented by D.W. Griffith in 1908. Chung and his team also had to solve what he calls "the identity issue". Inside a VR experience, the audience has to know who they are in the VR world, what to do there and how to relate to the characters. "Scale is the solution. It makes tech more intimate and allows you to build a relationship with the characters, which are designed to be very fragile looking while at the same time being capable of conveying their humanity," Chung says.

Allumette is available free on Playstation VR, Oculus Rift, Steam VR, and Viveport.

allumette

PENROSE

It's About the Bandwidth, Stupid

Latency problems plague MMOGs. Put simply, all the avatars in an interaction move at different speeds, limited by the power of personal computers and their network connections. Putting several graphically sophisticated avatars in one place stresses the computing and graphics power of the servers and our PCs. Avatars lose sync. Some slow down, pixilate or just freeze. Voice communication becomes impossible due to dropout. For this reason, Second Life is far from a graphic wonderland. They need to keep it simple to help dampen latency. WoW and the plethora of MMORPGs have better control over this because most of their graphics are being run on their local CD drive. Second Life is ever changing, yes, but they don't have all the graphic upgrades and expansion packs that Blizzard rode to two billion dollars of sales, and a better user experience.

AltspaceVR

Over the past several years, a number of social worlds have come online. Of course the granddaddy of them all is Second Life. But since 2015, a number of social worlds have opened their virtual doors. The first, AltSpaceVR, was one of the VR ecosystem's most promising companies until it stumbled in August 2017, announcing abruptly that it had failed to secure more financing and would be closing its doors. Two weeks later AltSpace was acquired by Microsoft. "It was a question of timing," Eric Romo, AltSpaceVR's co-founder and CEO, now a Microsoft executive, told me in an interview. Romo attributes the company's difficulties to the slow adoption of VR. "There are 60 million VR-ready PS4s in the market, yet Sony has sold less than two million headsets. The value proposition is clearly not enough to convince gamers to switch. We still live in a world where only maybe a million people a month put on a VR headset. The masses may not be here yet, but we've got a huge head start with a community of 30,000 dedicated repeat users."

"Because people's social graph is not online," Romo continued, "social discovery is a big deal right now." To help break the ice AltSpaceVR is adding a suite of social games. The company has a free open source SDK to make it easy for developers to create or import games into AltSpace. Romo says the resulting games, such as "Holograms Against Humanity" (derived from the popular party game "Cards Against Humanity"), and a YouTube Jukebox, have been a hit with users. Facebook Spaces is also adding casual games, and inviting third parties to add their own.

Sansar

Second Life's parent company, Linden Lab, recently launched a beta version of a new social VR world, Sansar. They built the platform from scratch, unhindered by legacy engineering. "Sansar democratizes social VR," says Altberg. "Until now, complexity and cost have limited who could create and publish in this medium. Sansar dramatically changes that." reators were nvited to use the Sansar platform during a limited access preview and have published thousands of amazing experiences. Sansar's atlas directory already features hundreds of engaging virtual experiences, including multiplayer games, recreations of historic sites and landmarks, art installations, movie theaters, museums, narrative experiences, jungle temples, 360° video domes, sci-fi themed hangouts, and more.

High Fidelity

Meanwhile, Linden Lab founder, Philip Rosedale, who left the board of the company he founded in 2013, has developed a new approach to social VR, High Fidelity (HF). "Sansar and Linden Lab (and VRChat, AltspaceVR and everyone else) are hosted services where they run your servers for you. This means they probably collect usage data (privacy issues), and may impose community/content guidelines that don't fit your use case. By comparison, High Fidelity makes open source software that you use on your own servers to create a VR space," said Rosedale in an email exchange. The company recently announced a turnkey partnership with Digital Ocean that makes it easy to start a private server. Plans start at $5/Month. The High Fidelity client is as easy to install as a browser, and today it runs on Oculus Rift, HTC Vive, and the new Microsoft mixed reality headsets. It can also be used in desktop mode without an HMD. In early 2018, High Fidelity plans to release a version for mobile devices.

Like Second Life, High Fidelity has developed its own cryptocurrency based on the blockchain. Things you buy can be used in High Fidelity and also other virtual worlds. "If Blockchain were around when we started Second Life, we would have built everything around it, and indeed that is some of the work

we're undertaking now with High Fidelity. I think that Blockchain is super important for money, for digital assets, and most importantly for identity. Not just in the virtual world, but in the real world," Rosedale concluded.

Rec Room

Rec Room is the most popular free to play social VR world right now. It's created by Against Gravity, a game developer and publisher based in Seattle, which recently received a second round of funding from VCs Sequoia, Vulcan and Betaworks. Rec Room features multiplayer games like laser tag, paintball and ping pong that are simple, intuitive and social. People know what to do there, and have a reason to interact with others over ping pong and similar social games.

In November 2017, James Bicknell gushed about Rec Room on VRHeads.com. "I have a new addiction in VR, and it caught me a little by surprise. Rec Room, a social VR app with quite a bit to do... it really blew me away." Currently there are six game rooms, with games like charades, ping pong, paintball, soccer, quests, and poker. "Private games are especially great for hosting younger kids so they can try to play without you worrying about swearing adults or any inappropriateness," Bicknell observed. Our editorial assistant recently received an Acer Microsoft MR VR headset. He reports his girlfriend was seeing lasers long after they put the headset away.

Oculus Rooms and Facebook Spaces

Oculus has had a social VR product for Gear VR, Oculus Rooms, which has been around since 2016 that allows users to create their own social space and invite friends, as they would with oth any social media network. AltspaceVR also has a Gear VR app, and mobile VR accounts for nearly half their audience.

Facebook Spaces is a social VR app developed by Facebook, starting on Oculus Rift, and it will soon come to other VR platforms. One of the Facebook Spaces personalization features allows users the ability to create an avatar that best matches their personality. Facebook selects photos from a user's Facebook profile. They then choose one and the system then provides five custom avatar choices that match it. Users can change any number of specific features, such as hairstyle or hair color on the avatar they have chosen. The resulting characters have a cartoony, almost South Park-like appeal to them. Together with friends in Facebook Spaces, users can use the VR selfie stick to shoot photos that can then be shared on Facebook, or use the live camera to broadcast live videos to Facebook.

Facebook recently released the first in an upcoming line of social mini games and experiences to be integrated into Spaces: Bait! Arctic Open that allows players to compete is an intuitive, easy to play, silly social game. These types of structured interactions are the most exciting recent innovations introduced in these early social VR platforms.

Social VR is a new form of social media that is just starting to find its audience. It's clear the industry recognizes that people need to do things in VR. If people do indeed prefer presence in VR to the relative simplicity of social media, social VR could be a killer app: A reason to buy hardware because of what it does. There is no doubt in my mind that when people start spending hours a day in VR, they will crave connections with others, and these kinds of social sites will gain more momentum.

AR Volumetric Telepresence Will Be Disruptive

Volumentic telepresence is going to be very disruptive, even to VR. This is where two people wearing AR headsets in remote parts of the world occupy each other's physical reality. This is becoming available today. It will soon penetrate businesses worldwide and allows global team presence and collaboration. It might take a while for AR to overcome the lead VR has in development. However, when people start doing telepresence at work, like email, they will want to do it at home.

"Art acts as cultural radar, an early alarm system for the development of media."

– Marshall McLuhan, Understanding Media

Artists Show Potential OF VR AND AR

by Charlie Fink

The development augmented and virtual reality (often referred to as mixed reality) is still in the "tools" stage. There isn't much in the way of content, as the tools to create those experiences are just reaching the hands of artists who make the magic.

ARKit and ARCore allow existing smartphones to do more things, but mobile AR is far from the game changer many said it would be. Snapchat Filters and Pokémon Go are its best known applications. VR, meanwhile, is dominated by primitive systems such as Google Cardboard and Samsung Gear VR (they represent about 65% of consumer VR devices), whose principal content is low resolution 360 videos.

Within these limitations, artists are using the tools at hand to create compelling experiences which foreshadow the enhanced digital world about to unfold. The VR Society, a non-profit organization created by major Hollywood studios and top technology companies to advance the arts of VR and AR,

Artist Zenka shows off her AR prints at The Art of VR at Sothebys, New York City. May, 2017.

recently presented "The Art of AR" at Sotheby's New York in June 2017. The two day conference featured technology demonstrations, experiences, and entertainment exhibits spread across several floors of galleries. Upstairs, the VR Society presented panels in the main hall where normally sheiks and moguls bid ten of millions for the world's most sought after fine art. Here's what struck me: Art and technology were the most interesting part. The commercial applications, which included VR experiences based on the Alien and Spider-Man movies were less exciting. It is great to see big media companies investing in this new medium, but their goal is hardly art. It's the expansion of their most important franchises. By the way, the value of movie franchises to virtual reality cannot be understated. People know where they are and what they want to do there.

The Virtual Reality Company (VRC), a content studio and production company founded in 2014 by Hollywood veterans including two time Oscar winner Robert Stromberg (Production Designer of Avatar), directed "The Martian VR Experience" for 20th Century Fox Innovation Labs in 2016. At Art of VR, VRC occupied a large gallery which they filled with production art and a dozen seated VR stations to present its newest immersive production, a beautifully animated short, "Raising a Ruckus". You have great presence inside the world of the narrative, though no agency to move or affect the action yourself. You're the camera on a roller coaster ride of tracking shots and swirling effects. "We are in the early days of unlocking the power of this new medium," said Stromberg. "It's similar to my time working on Avatar,

VRC is in a constant state of invention and innovation, testing and pushing the artistic, storytelling and technical potential of VR."

Perhaps the most interesting work came from 2D artists using new media to digitally augment their physical work. Sculptor and printmaker Zenka, whose work is in Appendix 1 in this book, creates black and white lithographs that are augmented when viewed through our Fink Metaverse app. The prints themselves are the triggers.

Jane Lafarge Hamill is best known for her abstract oil paintings. She's represented by prestigious New York art gallery FMLY. Lately, she's been working in virtual reality, and not for money, marketing, fame, or any mundane reason. "I simply want to walk into one of my paintings," Hamill says. The viewer's entry begins before they even put on the headset, by facing the physical paintings across the room. Then, headset on, they see a scan of the same piece at the same size. "It's important to me that the viewer begins with a real object; having the actual painting there before they put the headset on and see its simulacra, is a subtle but integral part. This grounds it in reality." The viewer then walks through the surface of the painting, which deconstructs around you.

A friend gave Hamill a crash course in Unity. "It's a steep learning curve, and I've only brushed the surface." This tells us two things. First, you can create meaningful content on Unity, even if you only have an art degree and, most importantly, two, artists like Hammil are going to change the way we think about this new media.

Kevin Mack is an artist and visual effects pioneer who won an Oscar for his effects work on the movie What Dreams May Come. Mack has been nursing visions for immersive virtual reality art experiences for decades. At Sotheby's he was exhibiting his constantly evolving VR world, "Blortasia", which allows users to float or jet through a seemingly organic, ever-evolving lava lamp like environment. The experience is trippy, relaxing, weightless, and disorienting, all at the same time. "The potential of virtual reality as an art form is mind boggling. I've always dreamed of using technology to transport people to places and experiences beyond imagination. Virtual reality is making my life long dream an actual reality," said Mack.

Virtual reality galleries within the gallery were the subject of several exhibits. Notably, The Apollo Museum and the Harold Lloyd Stereoscopic Museum, were both built using Linden Lab's new VR platform, Sansar. They also run the enormously profitable MMO Second Life. Bjorn Laurin, VP of Linden Labs, told me Sansar will be the WordPress of VR. A simple world construction tool and platform that anyone can use to create their own VR world. In the Apollo Museum, visitors are greeted by a holographic Buzz Aldrin and can walk alongside a full size model of the Saturn V rocket. Silent movie legend Harold

Lloyd was enamored with stereographic photography for decades and took thousands of 3D photos of celebrities like Marilyn Monroe, and events like the opening of Disneyland. Lloyd's photos and other memorabilia from his long career have been lovingly curated by his granddaughter and placed in a VR museum that allows you to navigate as if in a real building.

VR and AR art will hardly be confined to galleries, and ultimately will be integrated into many of the things we view casually every day, like books, newspapers, billboards, entertainment, advertising, sports, and medicine. Even theater, among the most ancient arts, is embracing this new medium. England's Royal Shakespeare Company's recent production of The Tempest featured a digital Ariel, created using an on stage actor in a motion capture suit.

Case Western Reserve University in Cleveland, Ohio, has a unique relationship with Microsoft and has been a test bed for medical and design applications through its "think[box] Innovation Center". In November 2017, they produced an augmented dance performance, Imagined Odyssey. Audience members were given a HoloLens which allowed them to see the dancers interacting with digital images, at one point literally controlling a whirlwind. "I have always had a keen interest in the 'atypical' application of certain technologies especially in dance. So, when I created this project, there was so much to explore, consider, research, etc. My personal philosophy over the past fifteen years of this line of work has been that technology should be 'invisible'. The technology should serve the needs of the art, and not the art used as a means to showcase the technology," Director Gary Galbraith told me. "The use of this AR technology provided for a unique opportunity to expand the tool kit and explore a greater range of our art form in its stage presentation."

Kevin Mack, a VR experience designer who won an Oscar for his effects work on the movie *What Dreams May Come*, created the constantly evolving VR world, *Blortasia*, which allows users to float or jet through a seemingly organic, ever-evolving lava lamp like environment.

THEATER and AR and VR

By Tim Kashani

Tim Kashani is founder Apples and Oranges Studios/ ARTS and IT Mentors. He spends his time educating in technology and the arts, developing new innovations, writing, directing and producing for Broadway and film, and spending family time.

The art world has long functioned as a laboratory for utopian new social frameworks, yet-to-be assimilated technology, and new modes of storytelling. The explosion of new media in the arts has signaled a profound collision right now between technology and art. Artists are interested in testing the waters and prototyping new ways of seeing with new tools. This is all evidenced by the increasingly prevalent presence of augmented reality technology in everything from Broadway productions to video games, Hollywood movies to the contemporary art world.

Through its special developmental relationship with Microsoft, Case Western Reserve University in Cleveland was able to marshal over 100 HoloLenses for an experimental dance performance in November, 2017.

Story Immersion

The conversation about pre-show technology applications is perhaps most salient when it comes to innovative new branding strategies. With 360 degree video now hosted on YouTube, storytellers from a variety of production formats are starting to experiment, and find success with including an audience in the pre-show journey through immersive video teasers. This is where the lean startup mentality most shines. In order to move away from the ultra linear Broadway or bust model that characterizes traditional theatre productions, such applications of 360 degree video allows storytellers to bring audiences in and iteratively troubleshoot. 360 degree video is a great medium for gauging audience engagement as well as audience demographic.

The goal is to apply immersive marketing to generate committed fandom and emotional investment early on. Using storytelling devices like origin stories, behind-the-scenes reality tv with actors, and immersive teasers allows productions to generate a full picture of their target audience and a sense of prevailing preferences and ambivalences. 360 degree video is a powerful tool here, and it's evolving in its breadth. We can see all of this at work in the 360 degree video teaser made by Hamilton, in which an exclusive sneak preview of Hamilton was available in 3D in preparation for the 70th Annual Tony Awards. It goes without saying that marketing for Hamilton has been wildly successful, and its innovative technological applications are what's given it an edge over other productions. Earlier on, School of Rock: The Musical produced 360 degree video teasers, and aired them on YouTube. Their orientation placed viewers in the midst of the band, creating immersive engagement, and effectively building hype for what would be a notably successful production.

There are now two styles of 360 degree filming. One, where the camera is in the center of the action. Which creates a gamified environment that the spectator is in the midst. The other, where the camera instead circles around the environment, creating an amped up traditional theatre experience in the round. This second mode puts viewers on the outside, but also endows them with agency over where in the theatrical environment they'd like to explore. Intel is working on this new form of 360 filming, and its application promises to be unique. As of now, this new 360 degree filming style has been applied in major league baseball, which has set a pact with Intel to live stream games. Again, this represents a push into enhanced accessibility, with new demographics suddenly able to access a once high cost of attending a professional sports game live.

The reality is, in our current landscape, there's a diverse pool of storytelling technologies. Some of them online. Some of them offline. And some of them hybrid. Storytellers need to strategically be using technology to turn their concepts into a brand that can gracefully navigate from one format to another, and one technology to another. Once a production tips the scale and

becomes a brand, a story will market itself through the greatest and oldest technology of all: Word of mouth.

The Royal Shakespeare Company partnered with Intel for their 2016 production of *The Tempest* in London. Intel used twenty-seven projections to augment Ariel, a spirit which flies over the audience during the show, no AR glasses required.

Photo courtesy Intel

VR Set Design

On the side of creative production, virtual and augmented reality is showing immense potential as a tool for designing and troubleshooting environments: Sets, stages, and spaces where entertainment takes place. My company, Apples and Oranges, recently collaborated with Microsoft HoloLens to build an augmented reality set design tool. Built using the Unity game engine, the tool allows directors and designers to plan and design sets using mixed reality technology. It's a voice responsive AR diorama tool that lets users control set design, scale, lighting and angles in a virtual theatre space that can be beamed anywhere in the world. This has potential for international collaboration. Directors and designers in disparate locations can collaborate and evolve concepts, unbound by geography within their shared virtual project space. The platform uses algorithms to accurately scale allowing designers to virtually foresee technical issues, which has major money saving potential. And it incorporates gestural manipulation and voice command to change elements like lighting and sound. When companies design Broadway sets, they're still largely using dioramas not unlike the ones that were made twenty years ago. Essentially miniature set models that can cost as much as $20,000. The augmented reality set design tool alleviates this clunky problem by creating a virtual set design suite that can be worked on and prototyped in a cost efficient and collaboration friendly way. Moreover, exciting new possibilities for rehearsal and story prototyping are bound up in this new technology. With a virtual set or space, actors can have physical access to set pieces during rehearsal before the set pieces have even been constructed. This potential can be mapped onto escape room type theatre productions, which can benefit from virtual construction in early concept phases.

Post Show Storytelling

So far, we've dived into the multifaceted applications of new technology for pre-show branding, production, and dissemination. The 'post' element is also charged with potential. All of the aforementioned tools shouldn't be put away after the show has finished running its course. They all have specific capacities to keep a story relevant, sustain engagement, and lay the foundation for future iterations. The beginning of a franchise, or a re-imagining of a narrative in a new medium.

The virtual set building technology discussed above has interesting potential to be used for post show revivals and theatrical exhibitions. These post show applications of technology are modes of evangelizing a show. Long after an audience has enjoyed a production, this tool will enable them to put on an augmented reality headset, and walk around and sit in the set pieces they once witnessed on stage. There's a prospect here for creating interactive retrospectives, and post show storytelling. Similarly, 360 degree video can be a powerful tool for transitioning a theatrical or live mixed reality performance into a web series. Creating a hybrid post show legacy, somewhere in between theatre, television, and gaming. The task of a production team is to keep an audience invested in a storyline in a way that sustains hype for a second related rollout of a story in a new medium. Post show technology application is rooted in the task of sustaining engagement even after closing night, in order to keep options open for breathing new life into a story in different formats down the road.

During The Show

During show technology applications are the most spectacular, and are increasingly viable for even small-scale theater companies. What follows are a few especially notable examples, which have set the tone for experimenting and pushing boundaries on stage. The Royal Shakespeare Company partnered with Intel for their 2016 production of The Tempest in London. This production used performance capture technology to bring digital avatars to stage in real time, based on facial and movement capture technology interfacing with real actors. Intel used twenty-seven projections to augment a character named Ariel who flies over the audience during the show. I's an example of during show augmented reality that doesn't require the audience to use any glasses.

Another recent example of theater implementing these technologies during-show is the Broadway production of The Lion King. It marks the first time anyone's recorded a live show on Broadway, with its 360 degree rendition of "The Circle of Life." This can be viewed on Google Cardboard via YouTube 360, an Android device, or on a Samsung Gear VR headset via MilkVR.

Moreover, technologies that are slightly older than 360 degree filming and augmented reality hold immense potential for creating immersive and active theatre experiences. The immersive Michael Jackson Journey produced by Cirque Du Soleil uses surround sound, projections, and holograms to create an in-the-round world that audience members are at the center. The shift we're looking at is at much about new technology as it is about shifted conceptions of what constitutes viewership, and what entertainment looks like. Currently, my company is coalescing all of the aforementioned technological applications in a new show called Higher Education. The production is putting to the test the range of non-traditional techniques. It's a prototype of theater of the future, and it's including the public from the early stages of its inception. Virtual reality technology will be applied for dance sequences, as well as 180 degree filming technology to create a hybrid theater film experience. As Higher Education develops these hybrid theater immersion forms, we'll be sharing everything through social media. The idea being to put out new methods as they're developed in order to prototype, revise, and then relaunch.

This all can be distilled in a maxim that surely everyone can get behind: "Taking the starving out of artist." With new technology on the horizon, and a rapidly evolving contemporary landscape that's blurring boundaries between theater, film, gaming, and social media, the theater world is opening up to new players and rule breakers, and there's more potential than ever for theater and performance to sustain a livelihood for artists and storytellers. And the best part? These spectacular immersive technologies are becoming increasingly viable on smaller budgets and with fewer resources. There's an overall ethos of open sourcing and networked sharing, pointing toward a future art world where gatekeepers are a relic of the past, and walls between audience member, actor, and artist are excitingly malleable.

Wearing a Vive VR Headset and standing in a subway car set, the user experiences a virtual world perfectly mapped onto ours.

A New Artistic Medium

Blackout uses a volumetric method of photographing someone from different angles simultaneously to create a 3D photo realistic avatar (albeit heavily pixelated) of a human being. In this case, sitting on a train. You enter a room which has floor to ceiling poles in it, like you might see in a real subway car.

Wearing a Vive headset, and using its room scale roaming feature, you can move around the car. When you reach out for a pole you see in the digital world, you touch a real pole. It's a very cool, convincing illusion. There are six people on the train. They cannot see or hear you. You're a ghost. As you near one of the riders, you hear their thoughts in their own words. A stream of consciousness internal dialog recorded in thirty minute interviews with the participants. The train doors open, and you get off on what appears to be a real platform. It is one of the most compelling VR experiences I've ever had.

I experienced Blackout at the Tribeca Film Festival's Immersive Arcade in March 2017. Somebody, somewhere, needs to restage this in an entertainment venue, or more appropriately, a modern art museum. It's one of the most compelling VR experiences I've ever had.

On ubiquitous visual computing, the death and rebirth of the smartphone, and the self driving car as web browser.

It's 2025. My son, Lucien, hops into an Uber headed for university, his master's program in robotics. A car that was designed completely in virtual reality; that was manufactured in a factory designed in virtual reality, and assembled using processes visualized in VR, by workers trained in VR simulations. A car that lives in a network of millions of self driving vehicles, monitored by operators using augmented reality dashboards.

To pass the time, Lucian checks his social feed via a beautifully rendered AR display that transforms the mostly empty cabin interior of the car into an interface to his digital life. Then he calls up a Twitch feed to check on the latest vSports tournaments, and ultimately settles in to binge season 10 of Friends, because he's feeling retro today. All of this is called up using a voice interface and explored with a combination of hand gestures and eye tracking.

Lucian's TV watching is interrupted by an incoming call. It's his mother Marina reminding him to eat. After a quick volumetric conversation with mom, he browses for food choices on Amazon, and redeems blockchain based coupons to pay for his lunch, which is waiting for him when he arrives at school.

And that's how my talk at the VR Strategy Conference kicked off. I had the distinct pleasure of sharing a stage with Rikard Steiber, president of Viveport and SVP of VR at HTC Vive, in a fireside chat about the future of immersive technology titled "VR 2025."

After swapping origin stories, we each shared a vision of the world in 2025, starting with Rikard. Rikard's world of the future is one in which we're no longer talking about VR or AR, they're just part of everyday life. VR and holographic displays will be commonplace, and much lighter and comfortable. Because much of the computing will be distributed in the cloud, or on devices that are decoupled from the display and untethered. Rikard dubbed this idea as the "death and rebirth of the smartphone." That is, mobile computing isn't going away. And in eight years, we'll still be doing all the things we do on our phones today like making calls, sending emails, and watching movies. But because of advances in displays, visual computing hardware and software, and 5G networking, we'll have a ubiquitous 3D infrastructure that allows us to put the processing power where it's needed, and deliver amazing immersive content everywhere, in a variety of form factors, for both consumer and business use.

Me being me, I hadn't actually prepared all of my talking points, I played off Rikard and spun the tale of the self driving car of 2025. On the spot. Yeah, I'm that good. But the thing is, it wasn't at all hard, because most of this is already happening. Big auto makers are experimenting with designing cars in VR, with the idea that someday we will dispense altogether with physical clay prototypes. Assembly plants are being visualized in VR, along with the procedures to do the assembly in VR training modules. And 3D physics engines are being used to perform millions of simulations per day to ensure safety before putting all those hunks of autonomous metal on the roadways.

Once our cars can drive themselves, what are we going to do in them on long rides? I already check my Twitter a hundred times on average during my Uber ride home. But when these are self driving, and we don't need a front seat for the driver, what then? Auto manufacturers are already contemplating a redesign that includes in-vehicle entertainment centers that deliver a full menu of content. That will certainly start with flat screens, but as XR technology miniaturizes, and the other trends Rikard cited continue, it won't be long before those entertainment centers become fully immersive, and offer a vast array of additional services, including e-commerce, social media, web search, and why not phone calls? Le phone est mort, vive le phone!

What could that look like? Will it be a Vive in a car? Probably not. In air holograms? A lot can happen in eight years but that seems like a stretch. We're probably talking a pair of smart glasses, per Rikard, wirelessly connected to an onboard device that does some computing locally, but goes to the cloud for the hard stuff.

Some kids are already working on the design for all this.

To me, it doesn't seem like too far fetched of a vision for 2025.

And that's just cars. **– Tony Parisi**

THE NOT IMPOSSIBLE PATH TO REALISTIC VR

By Charlie Fink

In virtual reality, the more realistic the digital world, the greater the demand for fidelity. The greater demand for fidelity, the greater the demand for bandwidth. Nothing is more demanding than cinematic realism. We expect photo real worlds to have absolutely perfect fidelity. This is why realism in VR always involves a sterile space station or a desert planet. It is going to be some time before we cavort with artificially intelligent characters in Westworld or The Matrix.

Augmented reality will soon be more than cartoony Pokémon and Snapchat Filters, and Facebook's forth coming camera effects platform. It will soon feature realistic holograms of people. "VR has given birth to a process at the intersection of games and movies," said James George, co-founder of DepthKit.

3D capture is accomplished by a process called photogrammetry, which uses multiple cameras to capture high resolution images of a person or a place from as many angles as possible. And then, using a computer program, pieces together all the images into a 3D person or environment. Unlike 2D cinematography, the processes for capturing the environment and people in 3D for VR are separate. Developers must place the realistic 3D characters into the world they created. Using VR, you could then enter that world and walk around with them. Of course, you'd be a ghost. You wouldn't be able to affect the characters of the action.

Using DepthKit, anyone can now do volumetric capture using a basic SLR, an Xbox Kinect, and a high end PC.

A multiple character drama in VR or AR would be like Sleep No More, the smash hit environmental theater production in New York, now going into its sixth year. It might also be so bandwidth intensive, it would take many hours, even days, to download over a typical home cable modem. Last June, I met Pete Forde of itsme at the VR Toronto conference, where he did a realistic 3D capture of me, and then animated it. Seventy cameras captured me as I stood, arms akimbo, in a heavily lit booth in itsme's office. The image was processed automatically in just a few minutes.

My animated avatar can be placed in any VR or AR application. Forde says he's working on a 3D capture method that will do away with the booth.

To promote the film Marjorie Prime, volumetric capture startup 8i captured star Jon Hamm (an AI character in the movie) and has just now made him available to everyone via their free Holo smartphone app. 8i previously unveiled the Jon Hamm hologram at the 2017 Sundance International Film Festival. Using 8i's Holo app, users can place the Holo-Hamm in their living room, snap a selfie, or record a video. The Holo app features a variety of photorealistic 3D holograms that can be added to your camera view, moved, rotated and resized. Holo unveiled a similar collaboration for Spider-Man earlier this year.

Holo-Hamm and Spider-Man are grounded in place. In the theater, that's called "park and bark". Capture stages are small. A tracking shot with forty-one cameras is hardly feasible. 8i CEO Steve Raymond told me the company's overarching goal is to "capture human performance so authentic they bridge the uncanny valley". Uncanny valley describes the hypothesis that human replicas which appear almost, but not exactly, like real human beings elicit feelings of eeriness and revulsion (or uncanniness) among some observers.

"The process of capturing a person volumetrically in 3D from an array of cameras pointing inward is called volumetric capture. Volumetric capture is video based, which makes the approach photoreal and scalable. It allows us to reconstruct the person being recorded with full volume and depth, color and light, so that when viewed in a VR or AR headset, or in mobile AR, the person looks real as if they are actually there and you can walk around them and see them from any angle," 8i spokesperson Amy Sezak told me in an email.

While capturing multiple photo real 3D characters in dramatic situations is nearly impossible today, the photogrammetric capture of the real world is relatively easy.

realityvirtual of New Zealand is developing a method of VR cinematography they call volumetric videogrammertry. They use photogrammetry, drones, and high speed cameras to capture real environments in startling detail. Because the file sizes are enormous, the company has developed a way to reduce the files

PC and console games like *Grand Theft Auto* and *Battlefield 4* create graphics that nearly achieve photorealism with motion captured characters in beautifully rendered 3D worlds.

automatically, without sacrificing fidelity. Founder Simon Che de Boer told me his company is creating "a one button cloud based processing service for people who want to make cinematic VR".

Like itsme and realityvirtual, DepthKit is a seed stage startup that provides the tools independent content creators need to capture volumetric video of actors to tell their stories in VR. DepthKit is a suite of tools that allows creators to capture, edit, and publish volumetric experiences. To use DepthKit, all that's needed is a camera, a Kinect, and a suitable computer.

DepthKit's method is the most inexpensive, and artists are able to stylize the results to go beyond just photorealistic looking holograms. People can be made more impressionistic, and therefore less computationally demanding. "You don't need to capture reality in perfect detail to make to make a movie believable. Movies exist in a kind of magical reality already," said co-founder Alexander Porter.

Co-founders Porter and James George developed DepthKit for projects they were producing themselves through their other company, Scatter. They decided to make the unique methods of low cost volumetric capture they pioneered available to everyone.

DepthKit supports the export of an image sequence for later use in standard editing applications like Adobe Premiere, Final Cut Pro, etc.

On the other end of the spectrum are the high end Framestore and Quantum Capture. Quantum, based in Toronto with a Los Angeles office, has a method that captures and then animates avatars. They are marketing this to both entertainment and game producers. "Our characters are completely naturalistic," CTO and co-founder Craig Alguire told me. "Facial animation and lip sync are perfect. These are essentially programmable humans."

In 2013, Los Angeles based Framestore brought Audrey Hepburn back from the dead for a chocolate commercial. Hepburn's re-appearance generated some nice royalties (aka "found money") for her estate. Of course, reanimation requires skilled animators, but animators are the most versatile actors in the world. A lion cub one day, Maleficent the next.

A bootstrapped Israeli startup, Octosense, also specializes in capturing performers in 3D and realistically animating them. According to CTO Omer Breiner, the company "makes high level realistic 3D content with files sizes no larger than video". Octosense is positioning itself to sell B-to-B services to media companies. Breiner thinks their technology could greatly enhance a digital assistant.

In 2013, LA-based Framestore used modified motion capture to covincingly reanimate a 20 year old Audrey Hepburn for a chocolate commercial.

"Major advances in mixed and augmented reality always seemed unachievable," said Octosense CEO Deve Ouazanan. "We've already made amazing progress and are now working to release the first realistic personal digital assistant. Our R&D department is already setting its sights on what can be achieved in terms of realism and natural facial movement, as we create the next generation of mixed reality."

The tools to capture and create realistic performances for VR and AR are just now reaching the hands of creators. This will greatly enhance the artistic and storytelling potential of this new medium. Even with the crudest of tools, it feels like a door to a new world is opening. ■

a new MOVIE

By Charlie Fink

"The movie business is in trouble," a senior new media executive at a major studio confided in hushed tones at the "VR on The Lot" conference last October. The conference was a VR/AR education event for the Hollywood community organized by The VR Society, a non-profit founded by leading entertainment, game and technology companies. "Attendance is crashing. Don't quote me on that. The quality of in-home is so good today, the challenge is to offer something they will never have at home." A long line of speakers confidently predicted that Location Based VR (LBVR) was going to be one of the key economic drivers of this first wave of consumer VR products, while exposing countless millions to the technology.

There are several companies working to expand the appeal of traditional movie going with new technology. The highly immersive Barco Escape three screen system is the most successful of these and boasts of a remarkable 62% boost in box office for Escape showings over other formats. Last year's

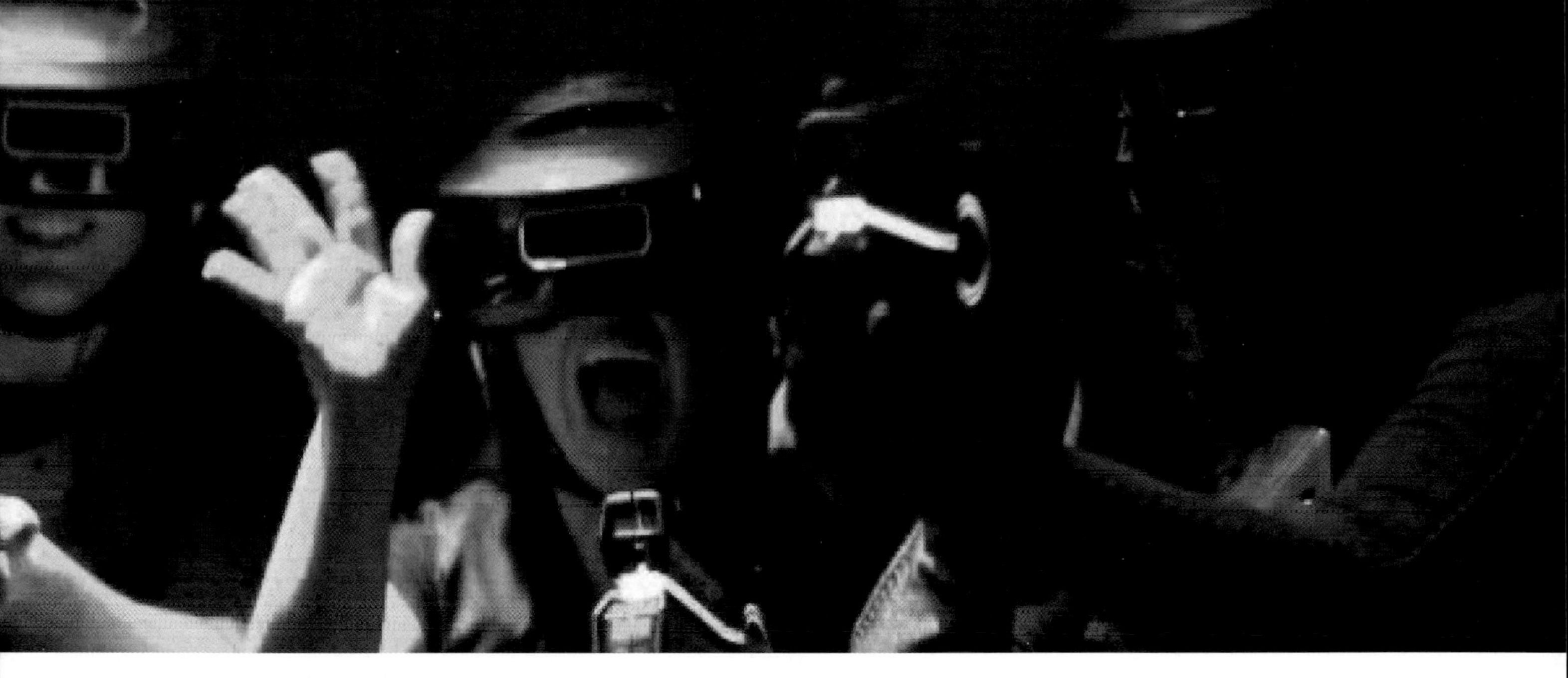

Friends suited up and ready to enter The VOID at Madame Tussauds in New York, which features a free roam *Ghostbusters* experience.

Star Trek release used the system to great effect. Donald Fox, a theater operator with locations in Pennsylvania and Maryland, agrees that the way to get more people in movie theaters is to make them better. But he's taking a slightly different approach.

Fox Theaters recently installed luxury recliner seating, and added a full kitchen and bar. "We reduced the number of seats by 50%, and increased admissions three fold," Fox told me. "Reserved premium seating with made to order food and cocktails makes people more comfortable, engaged and relaxed. As a result, they like the movies more."

Phil Contrino of The National Association of Theater Owners said Fox's strategy is part of a trend within the industry to upgrade seating and other amenities, such as dated arcades. "Movie theaters have somehow been labeled as a legacy industry that is very static, and that's just not true. Some of the most important innovations that have happened since the beginning of watching movies on a big screen have been happening recently." LBVR comes in four basic sizes:

- VRcades that feature head mounted displays (HMDs) from the HTC Vive high end home system (until recently Oculus did not permit public use of its equipment).
- Stand alone casual VR such as VR Coasters with a minimal footprint.
- VR theaters, which usually feature swivel pods.
- Free roam, or warehouse scale, VR.

The bulk of the activity is in China right now, with the US poised for dramatic growth. In 2016, China had an estimated 5,000 VRcades far outstripping any other nation. By contrast, there are perhaps 1,000 in North America and 500 in Japan, Korea and the rest of Asia. A headset and a laptop can pop up anywhere, so the actual number is tough to pinpoint. While only two fully themed "Viveland" attractions exist in major malls in China right now, many VRcades are in Internet cafes. However, unlike an Internet connected PC, each user needs an attendant to operate and instruct. VRcades are experimenting with memberships, subscriptions, or simply a flat rate per hour. Alvin Graylin,

VRStudios has over 40 free roam locations inside theme parks like Universal Studios Tour and Knott's Berry Farm.

We reduced the number of seats by 50%, and **increased admissions three fold**

President of HTC China, says "there's no clear data on exactly how many [VRcades] are operating. But it's true many who jumped into doing arcades without proper planning or resources have had a tough time."

One year ago, Springboard VR of Oklahoma City, Oklahoma opened its first VRcade based on the HTC Vive VR system. Founders Jordan Williams, Will Stackable and Brad Scoggin quickly realized the need to consolidate software operations and create a simple framework that allows users to make reservations and manage games. So they created the Springboard VRcade management system, featuring game preview, first time tutorial, time tracking, scheduling, content licensing, game launcher, quick setup, call for help and more, integrating seamlessly with the most popular POS (point of sale) systems. Their SaaS (software as a service) company now has 185 clients and is growing 25% every month. "It's the wild west out there," Williams

told me. "There are a few people with multiple locations, but mostly it's just people trying it out with a few Vives. One of the biggest challenges these arcade owners face is game licensing, which is one of the most important services we provide."

One step above the LBVRs supported by Springboard are the much hyped IMAX Experience Centers from the big screen giant. While superficially slicker than their regional counterparts, IMAX Centers are also based on the HTC Vive. They also offer the more immersive advanced Starbreeze headset at a slightly higher price. There is one IMAX VR center in LA, and one on the East side of Manhattan. The Manhattan location is almost always empty when I walk by. A manager told me they are busy on the weekends, but they have throughput issues. With a maximum forty-eight turns per hour, people have to wait. Most say they will come back and never do. Four new IMAX Experience Centers are planned for next year. The company has set aside a $25 million fund for the creation of new VR content, though it continues to characterize its VRcades as an experiment.

Warehouse scale VR is as close as one can come to walking around inside another world. Using a backpack PC, headset, and carrying props, guns or tools, you enter the simulation and become someone else, in a fully digital world. Honestly, it is the only true VR. Even the weakest of the offerings is down right amazing and worth every penny. However, repeatability is an issue. At a million dollars a title, one wonders how often will they be able to update. Some experiences are built as cliffhangers, encouraging the purchase of the sequels. Some seek to build in levels, allowing users to gain mastery, and planting easter eggs which can be revealed in each subsequent immersion. The environment reacts to the participants, adding yet more variation.

As many as sixteen people can share some simulations, relieving throughput pressures at peak times. Warehouse scale is also by far the most expensive. Several dozen powerful and expensive trackers are needed. The backpack PCs communicate with a central server. A Zero Latency partner told me the system cost half a million dollars to install.

The free roam providers in the market (or soon to be) include The Void, Zero Latency, Dreamscape, VRstudios, Nomadic, and Tick Tock Unlock. Each is taking a slightly different approach to the technical and business challenges facing LBVR.

The Void, of Salt Lake City, UT, has used its first mover advantage in the U.S. to position itself on the high end, opening Ghostbusters at Madame Tussauds in New York in the summer of 2016. Last fall, The Void was named one of eleven companies in Disney's prestigious entertainment technology accelerator program. The result is a deal to open two locations, Disney Springs in Orlando, FL, and Downtown Disney in Anaheim, CA, and feature a new

The new Star Wars free roam experience from The VOID, *Secrets of the Empire*, opened in November, 2017 to rave reviews.

Photo courtesy of Lucasfilm, ILMxLab, and The VOID

Star Wars adventure, Secrets of the Empire. Those are premium locations because of the constant, new entertainment seeking foot traffic. While the rent is set accordingly, it also means The Void can charge premium prices while maximizing utilization, minimizing marketing costs, and reducing the need to refresh titles as frequently. Secrets of the Empire, like Star Tours, could last a decade or more.

The other challenge to this approach is scale. There are a limited number of such high traffic tourist venues available worldwide. Of course, increasing your footprint in the Disney parks is scale, so the Disney relationship makes The Void the instant industry leader. Amusement parks are great venues for free roam VR, even as a premium ticket, as it is at Madame Tussauds.

The vision and technology for VRstudios began almost five years ago with three young entrepreneurs that set out to build the first large scale arena for competitive immersive VR experiences. This resulted in the first completely wireless multiplayer, room size systems that went to market in 2014, and were broadly positioned as a platform for all types of commercial enterprise applications. Kevin Vitale, the current CEO and seasoned technology executive, joined the company in 2016 and focused exclusively on LBVR.

Vitale and his partners have diverse backgrounds in enterprise technology, software, entertainment and game development. VRstudios is well known inside the insular theme park industry. Their clients include industry leaders like Universal Studios, where they offer a Halloween themed VR experience, part of their aptly named Halloween Horror Nights at Universal Studios at Knott's Berry Farm in California, which has its own VR experience for up to 16 players, Showdown in Ghost Town.

VRstudios also has installations in movie theaters, family entertainment centers (FECs) and casinos. Those are four minute experiences, which maximizes throughput. VR Studios recently released a twenty-five minute adventure Terminal 17, for it's new VRcade Arena large scale system. Eight visitors suit up in backpack PCs and strap on high resolution industry standard headsets in order to become a squad of galactic marines dispatched to clear a space station from alien bugs. The experience includes some cooperative escape room puzzle solving elements mixed in-between mayhem for some depth and variety. Like all good simulations, you lose track of time inside the bug infested futuristic power station. "We look at this from the point of view of the operator," says Vitale. "Our system is built to minimize costs and ensure ROI."

Modular free roam VRcade system WePlayVR by AiSolve requires only one operator and has a very small footprint.

Dreamscape Immersive recently made headlines with AMC's $20 million investment, a jolt of funding that hit before the company even had the chance to open its first location, slated for the newly renovated Westfields mall in West Los Angeles. The free roam LBVR venture, headed by former Disney Imagineering Chief Creative Officer Bruce Vaughn, is preparing a breathtaking new adventure for its opening. It consists of three ten minute experiences with cliffhanger endings. The company will open with several experiences. I was given an early demo of one of the titles when I visited their Los Angeles office and was absolutely stunned by an early walkthrough. They are keeping their debut titles a secret, so all I can say is that the new title is a great concept, exceptionally well executed with show business panache. Up to six visitors can share each experience. Like The Void, Dreamscape Immersive features modular set pieces, over which graphics are perfectly mapped. Railings in the digital world can be touched. A stick can be a flashlight, a torch or a wand. There is wind, and soon, smell. The cliffhanger ending will encourage guests to sign up for the sequel, which may also be ready at opening. Tickets will run around $15 for a ten minute experience, $20 at peak times. Dreamscape Immersive plans to offer packages and subscriptions which will encourage advance booking and repeat visits.

If quality factors into success, Dreamscape Immersive is well positioned. In addition to Vaughn, the company is led by Co-Chairman Kevin Wall, entrepreneur and producer of large scale events like Live 8, and Chairman Walter Parkes, producer of Men in Black, Gladiator, Minority Report and perhaps best known in geekland for writing the classic 1980s hacker movie War Games. Along with AMC Theaters, Dreamscape Immersive's investors include Warner Bros., 21st Century Fox, Metro-Goldwyn-Mayer (MGM), IMAX Corporation, Westfield Malls, and Steven Spielberg.

Zero Latency founder Tim Ruse and his partners have come remarkably close to realizing their ambition to a create a completely new form of interactive entertainment. The company tells me they opened the first free roam VR center in the world, predating even The Void. They currently have thirteen locations in Australia, Japan, Europe and the U.S. with area licenses in place with for ten locations in 2018.

"In the end, it's going to be a content game, just like computer games today. You have to develop hits. Those are the killer apps for VR," said Ruse. Zero Latency's current experiences are Zombie Survival, Engineerium, and the upcoming Outbreak Origins. The eight person survival game puts the guest inside a compound under assault by zombies. Ruse says all Zero Latency games engender competitive team play. For this reason, 30% of their business is repeat play. The other current offerings are much more ambitious and heady. Engineerium puts you in an Aztec themed floating M.C. Escher world, and suggests someone could make a substantial work of art using this new medium.

A new free roam VR company, Nomadic, came out of stealth at CinemaCon in March 2017. Founded by veterans of Industrial Light and Magic (ILM) and Electronic Arts, the company has not yet announced their first location, which they hope to open in the spring. I visited their San Rafael, California headquarters last August and had a chance to experience *Chicago*, a ten minute adventure cleverly designed to guide you through an urban landscape to stop a horde of drones from destroying the city.

I liked it so much I did it twice.

Chicago also incorporates real set pieces and props, wind, etc. which completes the illusion. It's quite effective. For several weeks last October, Nomadic was in-residence at the Technicolor Experience Center in Los Angeles to promote its capabilities to potential partners of all kinds: Studios, exhibitors, game developers, writers, directors, investors, and operators. The company is hoping to meet the competitive challenges of the business through partnerships in content and real estate. "To compete with online and bring people through the door, shopping centers have to be more entertainment focused," said Kalon Gutierrez, Nomadic's head of growth. "There's a huge opportunity here. It's not only the next generation of the movie business, but the mall business too."

For Ali Khan, a trading technology expert from London and his wife, Samrien, the path to VR started with the success of their "Tick-Tock Unlock" escape rooms in Manchester, Glasgow, Liverpool, and Leeds, into which they have sunk their life savings. Their games feature theatrical "game masters" who guide the audience into HTC Vive VR headsets, taking the escape room experience into another dimension, while creating context and a mission. Although their headsets are tethered to a computer, users have room size mobility, can pick up and manipulate objects and can see one another's avatars inside the layered, mixed reality world. Customers arrive at the venue for health and safety training at ENSO, a global energy company but end up joining a rebellion against ENSO. "What is unique is the scale of the story, which begins even before your visit, and the interaction of live actors is of utmost importance because you need to know what you are going to do there." Kahn is raising money to open a London location which will be much larger and feature additional levels and, of course, more VR. Tick Tock has created a social role-playing game, which instills a quest for mastery in players that engenders repeat visits.

Two Bit Circus, a 50,000 square foot retail entertainment attraction (they call it a "micro theme park"), is a different idea. Set to open in downtown Los Angeles this year, they will offer food, beverage, alcohol, and a wide variety of entertainment, including a robot bartender, VR, and escape rooms. The company says it's going to experiment with different ticket schemes. VR World in midtown Manhattan offers a variety of VR experiences for a cover charge, like a nightclub. They have a full bar. So far it seems the model is working for them. I am personally skeptical of combining VR, cocktails, and food. It adds more cost, complexity, and difficulty to something that is already difficult and complex.

One of those looking at the big picture of the out of home entertainment sector is industry veteran, Kevin Williams, who just organized The Future of

Imax's first VRcades feature hardware and software from the high end home system, the HTC Vive.

Immersive Leisure 2017 conference in Las Vegas. An ex-Walt Disney Imagineer who worked on developing the DisneyQuest in the 1990s, Williams told me "a little deja vu hits me when I get called in to consult on some of these projects. Seeing some ideas that we once saw in the 90's dusted off and presented as new. Great technology is not a business plan."

LBVR can make money in theme parks, as VRstudios is doing today. But VR's ability to do so in stand alone locations may be highly prescribed. There are only so many locations like Madame Tussaud's in Times Square, Universal Citywalk, Las Vegas or the Walk Disney Park adjacent to retail where the next two US Void locations will be. The Chinese are taking another approach, building an epic billion dollar VR theme park, set to open later this year.

As cool as it is, no one has figured out how to profitably deploy and scale VR in public places like malls and movie theaters. Low end operators with Vive setups seem to be successfully running their locations, sometimes using the hourly model, like Internet cafes. Serving advanced users in this way is a lot less labor intensive. However, these are lifestyle businesses that don't have large enough margins to scale. No one is getting rich that way.

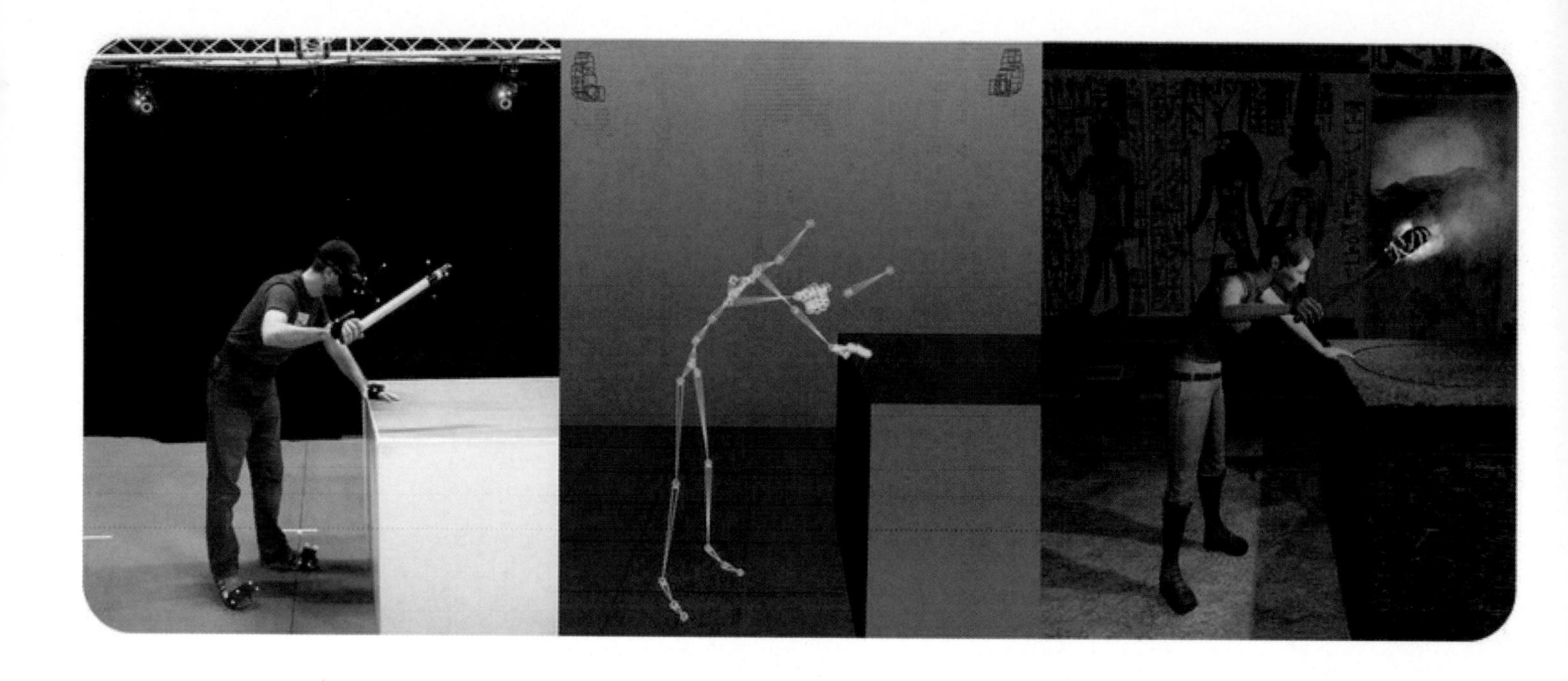

A deconstruction of how a scene is rendered in the Artanim system used by Dreamscape Immersive. Although not a single location has yet opened, the company, led by entertainment business insiders, has raised nearly fifty million dollars from media and real estate partners like AMC Theaters, Westfields Malls, and Warner Brothers.

This year, free roam VR companies Dreamscape Immersive, Zero Latency and Nomadic VR are going to forge beachheads in malls and movie theaters. Their free roam VR installations are nothing less than spectacular. Dreamscape may be able to charge as much as three dollars a minute at peak for an indescribable free roam VR experience made by some of Hollywood's greatest talents. Investors believe the product is so compelling to consumers it will overcome traditional retail barriers like marketing, labor (this is very labor intensive), throughput, and utilization (too few seats Saturday, too many seats the rest of the time). In my view, for these companies to succeed, they need to share the cost of developing content. Free roam experiences could be distributed like movies to venues like The Void, Zero Latency, Dreamscape, and Nomadic.

As much as I love warehouse scale VR and the wide breadth of potential experiences, I am skeptical about most business models outside of the general admission theme park. It's ridiculous to suggest that a few hundred people a weekend will change the fortunes of a struggling mall. It is far more likely VR will be dragged down with the rest of the mall's retailers. Nonetheless, panelists were certain of the future of out of home entertainment. I'm with Yogi Berra, who famously observed "If people don't want to come out to the ball park, nobody's gonna stop 'em." No one knows what's going to work, which is what makes this such a great story, still to unfold as I write this.

Perhaps there should be consumer VR (and maybe AR) centers in malls that demo both high and lower end VR. You could charge for the demos, deducting the hourly fee from the price of a new home system. There could be something akin to the Apple Store's Genius Bar for sales and customer service. The store could bundle PCs with HMDs, preconfigured, even preloaded, with VR software, making high-end VR as close to plug and play as it can possibly be.

The stores could preview new hardware like the upcoming HTC Vive Focus.

High-end home VR, even to consumers familiar with it, is still a consultative sale. If HTC, Oculus, and Microsoft were serious about exposing the public to VR, they would underwrite these VR stores as part of an integrated marketing strategy. I ran this idea past the Microsoft folks for their expanding chain of Microsoft Stores, and they said they already do this sort of thing. Though, that was not my recent experience in their flagship store in Tyson's Corner, Virginia. A Microsoft representative immediately contacted me to see what they do in New York. My proposal got an Oculus executive overly excited, but not in a good way. Oculus views the competition as the competition, not "coopetition," so alliances like this are unlikely. The good folks at HTC say they'd take something like this seriously if Microsoft and Oculus did. Nonetheless, I still think for a scrappy consumer electronics retail entrepreneur, this is a good play.■

The partners at Springboard VR started out as operators themselves. They now operate a comprehensive cloud based VRcade management system as a service to other operators.

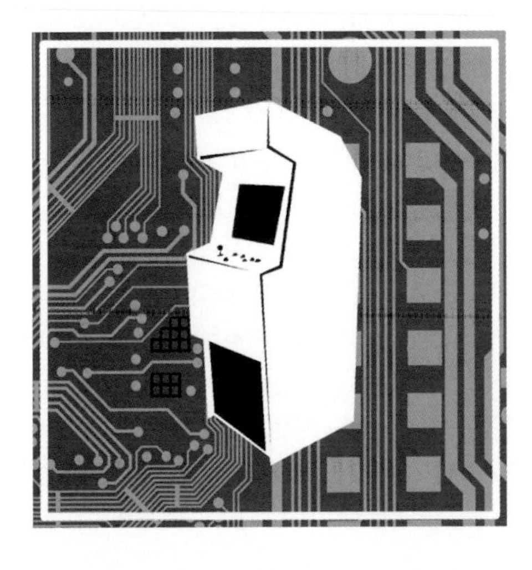

THE **AR/VR** ECOSYSTEM

By Charlie Fink

Inventing things is hard. Inventing many things and deploying them quickly is really, really hard. And sometimes out of reach for even the world's largest companies. This is because innovation and disruption are frequently accomplished by small companies, or small, independent teams inside large companies. Thinking different is often best done in an open, freewheeling environment. Also, some of the biggest breakthroughs have come from financially constrained inventors.

The current resurgence of VR technology was ignited by Facebook's $3 billion acquisition of Oculus in 2014. Famously, the startup was initially financed just a year earlier by a $2.4 million Kickstarter campaign. It was a textbook example of an industry leader using the strength of its stock price as the currency to purchase and scale a powerful new idea. Acquisition is the main way the majority of private technology companies will be monetized. Microsoft, for example, has been notoriously acquisitive over the past twenty

THE VR FUND H2 2017 VR INDUSTRY LANDSCAPE
APPLICATIONS/CONTENT
LOCATION BASED
SPORTS/LIVE EVENTS
SOCIAL
GAMES
ENTERTAINMENT
ENTERPRISE
HEALTHCARE
EDUCATION
TOURISM
SPORTS TRAINING
SOCIAL MEDIA
ADVERTISING & ANALYTICS
JOURNALISM
TOOLS/PLATFORM
DISTRIBUTION (APPS/MEDIA)
3D TOOLS (ENGINES/AUDIO)
REALITY CAPTURE (360 VIDEO/NEXT GEN)
INFRASTRUCTURE
HMD (TETHERED/MOBILE)
INPUT (HAND/EYE/WEARABLE/OMNI TREADMILLS/HAPTICS)
BY TIPATAT@THEVRFUND.COM

$2.5 Billion

AR/VR investment in 2017

Startup leaders across the twenty-seven AR/VR sectors raised $1 billion in the fourth quarter of 2017 as of this writing. This is only the second time the billion dollar figure has been reached in a single quarter. Three quarters of a billion dollars went into big deals like Magic Leap's $502 million and Niantic's $200 million, with VCs investing another quarter of a billion dollars into smaller rounds. AR/VR startups have raised $2.5 billion in 2017, equaling the record for AR/VR investment in a single 12 month period.

By sector, one third of all investment went into AR/VR tech since the start of the year. Just under a quarter of funds raised went into HMDs, primarily because of Magic Leap. Games took more than $1 of every $10 raised, with AR/VR navigation, photo/video, peripherals, entertainment and social startups also raising significant amounts. The remainder of investment was spread more broadly across startups in AR/VR advertising, art/design, business, e-commerce, education, enterprise/B2B, lifestyle, location based, medical, music, news, solutions/services, sports, travel/transport, utilities and VR headset sectors.

The range of VCs investing into AR/VR startups went from small, early stage specialist funds to global megafunds with billions under management. While AR/VR is still a very early stage market, the emergence of mobile AR and more advanced computer vision/machine learning has brought a broader range of investors to the space (even though my discussions show that AR/VR remains frontier tech in the minds of most VCs). The cooling of VC investment sentiment in VR in early 2017 has now been counterbalanced by AR (particularly mobile AR) focused thinking. **– Tim Merel**

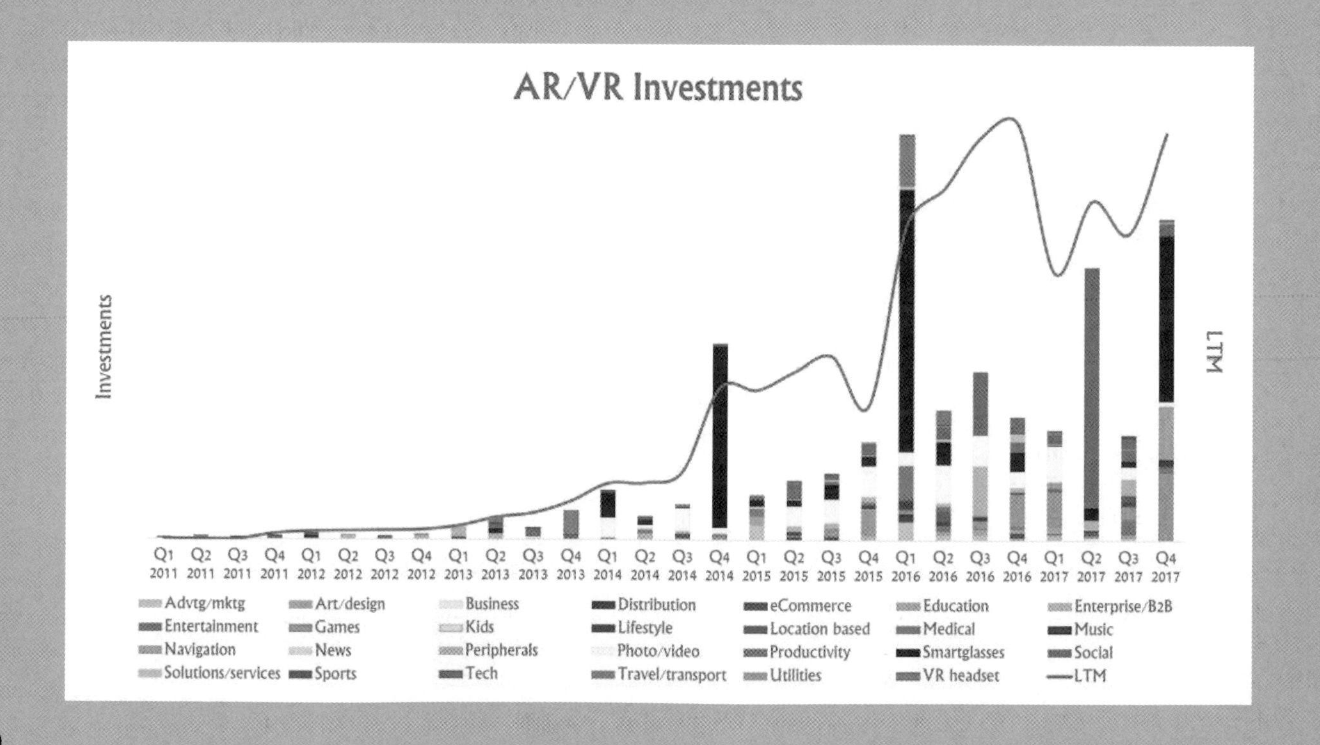

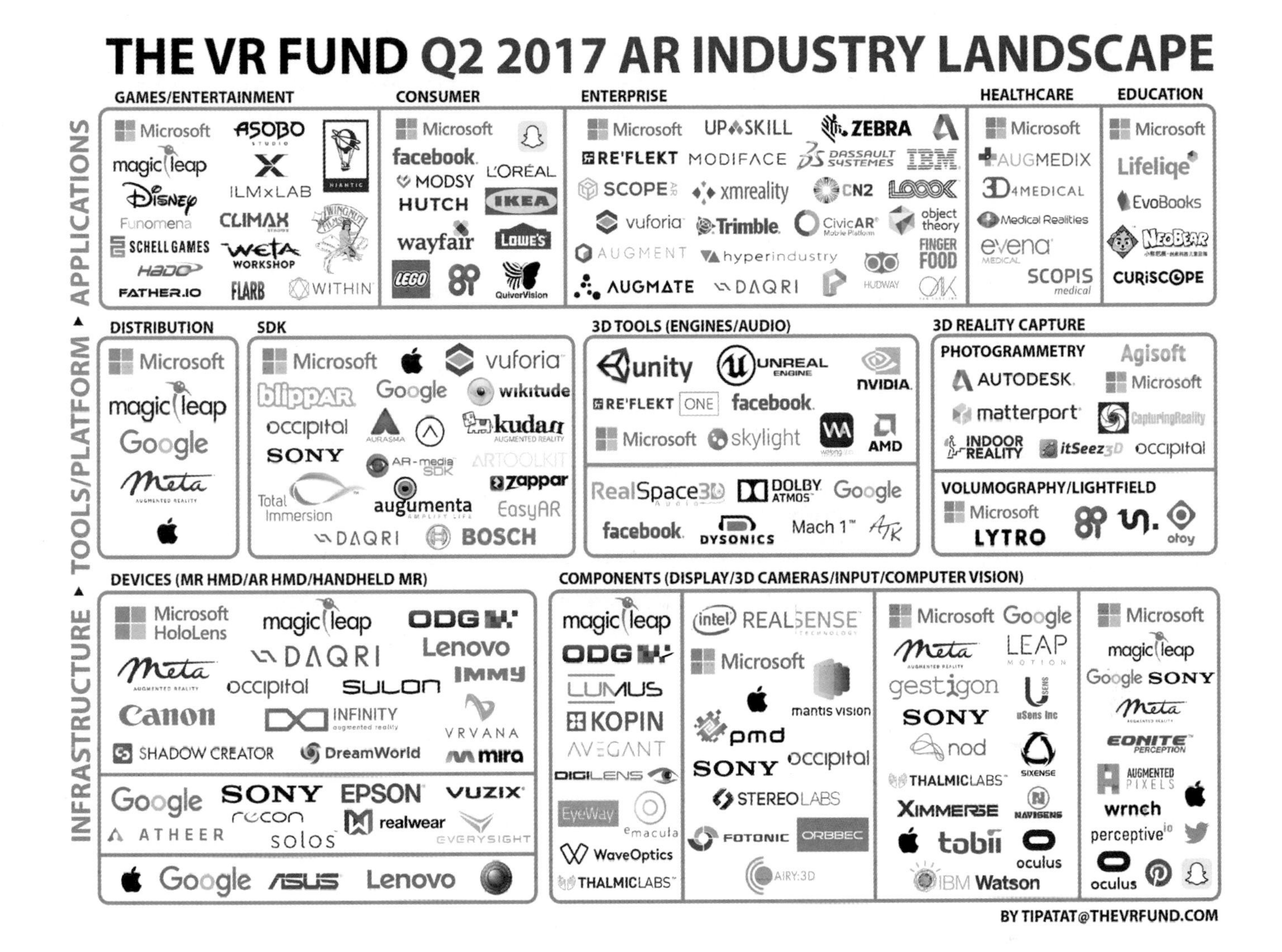

years. Swallowing huge companies like Nokia, and much smaller startups like AltspaceVR. Microsoft also recently acquired the popular cross platform game Minecraft, while Owlchemy, makers of the popular VR experience Job Simulator, was acquired by Google.

Over the past several years, a number of venture capital (VC) firms focused exclusively on AR and VR have emerged. The Venture Reality Fund (The VR Fund) with presence in Los Angeles and Silicon Valley, is focused on early stage investments, like Owlchemy Labs. An investment that gave the firm an immediate success when it was sold to Google last year. The VR Fund is committed to using education, information and community to strengthen the ecosystem, or environment where it is investing. They publish a semi-annual report and infographic to illustrate the size and shape of the VR and AR landscape, detailing the market's growth, new categories, new players, and merger activity.

Tipatat Chennavasin, co-founder and general partner of The VR Fund told me in an interview that "while talking to the community, I realized how little information there was out there, especially for companies to see what others are working on. People need to see both competitors and potential partners.

It's also important to allow people to compare and contrast companies at different parts of the ecosystem and understand what roles they play. We want to get new people entering the space up to speed as soon as possible".

"Already topping $2.3 billion in 2017, global investment in VR and AR is tracking higher than any previous period, and at a more accelerated pace," Chennavasin continued in an email. "The VR industry is healthy and growing steadily, and we are seeing increased investor interest from traditional venture funds, and new VR and AR focused funds alike."

The VR Fund evaluated more than 3,000 companies for inclusion in the 2H 2017 landscape and the 450 selected met a certain criteria of funding, revenue, mainstream coverage and/or major partnerships in categories as diverse as developing infrastructure, tools, platforms and applications. The free semi annual report highlighted these four insights:

The games category grew by 40 percent during this period. More than 35 VR game titles have generated $1M or more, signaling a healthy early stage ecosystem for VR game studios poised to grow dramatically in the next several years.

Notable global companies including Cisco, HP and Accenture entered the enterprise VR space during this period, but a majority of the category's growth was driven by startups building revenue generating businesses for specific industry sectors.

Major investment and growth within the next generation reality capture space indicates the broader market is shifting away from 360 degree video to true VR.

During this period, new all-in-one, standalone devices were announced, including the Oculus Go and HTC Vive Focus, which launched in China in December 2017. Priced at $500, the Focus may make a significant impact on consumer adoption in 2018.

The VR Fund views the ecosystem as a pyramid. Hardware infrastructure is at the bottom, content creation tools and distribution are in the middle. These enable the application and content built on top. "VR is more than gaming and entertainment and AR is more than enterprise," Chennavasin said. "With VR in particular, you can see that the bottom layer of hardware infrastructure is well established and most of the activity is in the applications and content layers. That means that VR is looking for the killer applications to drive mainstream adoption for both consumer and enterprise. AR on the other hand is where VR was a few years back. Still trying to figure out what the hardware infrastructure will be, before the developers can start focusing on the application and content layer."

Facebook has been the most public in terms of their support for the ecosystem. They recently committed an additional $250 million to fund early VR content, following another $250M invested in 2016. Sony, Facebook, Google, HTC, Valve, Samsung, Microsoft, Apple, Snap, Amazon, and others have been also very active in acquisitions and investments as well. In addition to the landscape infographic, The Venture Reality Fund hosts a Trello board with more information on the companies that comprise the ecosystem. The landscape graphics are updated twice a year, the Trello boards are updated weekly and show even more companies. ■

CONCLUSION

"It is not the strongest of the species that survives, nor the most intelligent that survives. It is the one that is the most adaptable to change." – Charles Darwin

As a futurist and technology columnist, I get to see everything and meet everyone, analyze collected information, and search for original insights. I talk to incredibly smart technology specialists who tutor me daily about what they are doing, and why it matters. Based on this information, and thirty-five years of experience in entertainment and technology, I form an opinion and off I go. It's a fairly informed opinion, which is nonetheless completely wrong way too often. If a futurist could foresee an iPhone or Facebook, he or she would invent it. I am first and foremost a social critic, telling one of the great business stories of our time.

While AR for enterprise is exploding and there is accelerated venture capital activity, VR sales to consumers was off in 2017, and AR is just getting started, mostly with nifty toys. The first adopters have adopted. Without lower cost VR capable PCs and an easily understood value proposition, VR will continue to be the purview of geeks, gamers and people who think Charlie Fink's Metaverse is a good title for a book.

VR and AR are not developing in a vacuum. As they mature, so will other technologies. Artificial intelligence, cloud computing, blockchain, improved wireless spectrum and bandwidth all have profound interconnection. There will be simultaneous advances in robotics and self driving vehicles.

Augmented reality is already making work more efficient by turning low skilled workers into high skilled workers, saving companies time and money. Offices will be semi-virtual, allowing teams around the world to collaborate as if they shared the same physical space. We'll move around the world much more freely, as Google Translate and other programs get integrated into more intimate devices, like glasses, or tiny hearing aids. Electric, self driving cars will be on the streets of major cities in the next ten years. Virtual reality is training surgeons and fast food workers. It's training pilots and professional athletes. VR is treating mental illnesses like PTSD and phobias. It relieves pain in hospitalized children. Artists have embraced it.

Killer apps are elusive, but not far away. Surely something social, like telepresence, will emerge as the compelling reason for the mass market to make the jump into VR. The AR Cloud will make AR ubiquitous and much more useful. This is all under development by companies big and small, and will be coming to market over the next decade. Telepresence, which allows multiple people to be volumetrically present anywhere in the world, is already being tested in the enterprise market.

The first adopters have adopted.

As technical obstacles like miniaturization and the AR Cloud are solved, science fiction will continue to evolve into science fact. Those who can afford it will live richer, more exciting lives. New forms of social entertainment will emerge. A massive multiplayer social VR universe, perhaps based on an existing franchise like World of Warcraft or more likely a variation on the infinite Star Wars universe, will makes billions. Users won't easily be able to tell humans and AI apart. It will be so awesome, people will choose it over real life.

Mary Shelley's 1818 novel, Frankenstein, tells the story of a brilliant young doctor, Victor Frankenstein, who miraculously reanimates a dead body. Frankenstein himself is immediately repelled by his unholy creation, but he cannot kill nor control it. The creature is intelligent and powerful, but also lonely and miserable. People see him as a monster, and he is shunned. His craving for love turns into vengeance as the monster uses the gift of life to torment Frankenstein and his family to death.

Along these same lines is the true-ish story of a brilliant Harvard undergraduate who started one of the world's biggest tech companies in his freshman dorm room in 2004.

As technical obstacles like miniaturization and the AR Cloud are solved, **science fiction will continue to evolve into science fact.**

The university traditionally provided names, headshots and contact information to facilitate the socialization of freshmen in the form of a book called "The Face Book." Mark Zuckerberg put the book online with a few social features which allowed users to "like" people, and get "poked" in return. Within two years, The Facebook was one of the world's most popular websites. Social media quickly joined email, e-commerce and search as the Internet's fourth killer app. Facebook makes many of the things we were already doing better, faster and cheaper. It is the ultimate address book and messaging platform. It unquestionably brings friends, families, classmates, colleagues and acquaintances closer together. It is deeply needed in modern, mobile times. We depend on it. But, to access these benefits, we willingly give away our privacy, which allows us to be segmented, analyzed, and influenced in insidious and subtle ways, of which we are not aware.

Facebook has had a profound effect on people all over the world, but there have been unforeseen consequences that Zuckerberg, like Frankenstein, has not been able to control. By his own account, Zuckerberg was unaware of secret political forces using the service to malevolent ends. The ability of Facebook to segment an audience into demographics and psychographics was heretofore considered one of the best things about it. At first, Facebook appeared to be a new town square where ideas and democracy could thrive. Instead, our new killer app may make us less free, and more vulnerable to bad actors.

Miraculous self driving cars and inventions like Elon Musk's ambitious hyperloop will change transportation as we know it, and will also have profound unintended consequences. Unless we are very careful, self driving cars and trucks will vaporize millions of jobs.

On the opposite side of the world, China seems to be planning on disruption, and is making a concerted national effort to re-balance the country's economy from one based on manufacturing to one diversified into digital goods and services. The Chinese are pushing VR use in schools, government and businesses, and the state is becoming a shrewd venture capitalist. China spends more on infrastructure than the US and Europe combined. China's Tsinghua University replaced MIT as the top rated engineering school in 2015. In 2016, China produced twice as many college graduates as the U.S. As Annika Steiber explained, the Chinese economy mixes potent features of a market economy with those of a planned economy, making it nimble enough to adapt to the demands and disruptions of our technology driven times.

I grew up in a world of landline phones, where computers and wireless phones were hard science fiction. In the course of my lifetime, one billion people in developing countries have risen out of poverty and joined the middle class. Smartphones, our small mobile personal computers, have given billions more access to all the world's knowledge, and now, new experiences with VR and AR. No one can know exactly how this amazing story ends, or if indeed it ends at all. The coming decade will bring more technological progress and disruption that we cannot fully imagine nor control. Of this we can be sure: Man's twin quests for immersion and augmentation will continue unabated.

Charlie Fink
New York City
December, 2017

Congratulations on finishing, well, starting Charlie Fink's never ending book on virtual and augmented reality. This book now has updated pages that are currently spilling into an online world. Pages that will have to be constantly updated in order to keep up with this new world of exponential change we now call home.

The fact that the book as a format is broken down on the side of the road is a big deal. According to Albert Allet Bartlett, human inability to understand the exponential function is the greatest shortcoming of the human race. What are the ramifications that a paper technology book has to be updated online nearly constantly?

The Changing Mindset Heidi and Alvin Toffler once said, "The illiterate of the 21st century will not be those who cannot read and write, but those who cannot learn, unlearn, and relearn." Fink has started us off with a deep, base understanding of this magical technology, but our job now is to keep asking questions, keep rethinking the answers, and then asking them again to see if the answers or questions have changed. When you combine AR with computer vision, blockchain, neuroscience, crowd participation, IoT, and 3D printing, the possibilities bifurcate exponentially again.

But please, relax. The first step to joining the future is to dump your mindset along with the immutable paper books onto the side of the road. Perhaps we have to get used to the fact that the world and the technology which is scaling up right now, is too complex to fully understand. Mastery is now a moving target, and that is why even the geeks-of-the-geeks, the computer programmers, have suddenly taken to a rare psychological infliction called "imposter syndrome".

Imposter syndrome is a phenomenon that causes people to feel like frauds. Inadequate in the face of their peers. Computer programmers often begin to specialize, and moving these programmers into another software stack or computer language can make them feel like their knowledge won't hold up. They panic that they will get "discovered" as a fake or incompetent in the computer genius world. But the truth of the matter is that all professions, from dentists to lawyers, are going to start to feel outstripped by this technological moving sidewalk that seems to be lurching forward and evolving at an overwhelming pace.

Moving away from mastery into curiosity could be the key. Just taking that step onto the moving sidewalk is important, and then while you are on it, look around, and learn what you need to learn to accomplish your specific goals, and then learn more around you all the time. The answers to the question you ask today about your AR headset, will likely have a different answer in two months.

Fear of not knowing, fear of the future, fear of incompetence, or fear of new technology is a fundamentally toxic brain state. Neuroscientists are starting to realize that fear is so bad for learning that they are proposing to make learning more like a game. When we relax and play, we can absorb information, experiment, progress and move forward.

Unfortunately, the other major roadblock between us and using all of these sci-fi tools and approaching possibilities is a harrowing sense of the "pressing present". We feel that the present is so urgent and all encompassing that we don't put any time away to experiment, learn and think about the future. This alone could us get left behind as this paradigm shift moves forward.

Time Bank Lifelong Learning Welcome to the 5 hour rule . I have yet to implement this successfully in my own life, but doing so requires nothing more than courage and opening up Google calendar and putting some non-negotiable time into your schedule. And in my case, it also means not scheduling meetings on top of it repeatedly and then giving up.

This is what famous CEO's like Bill Gates, Warren Buffett and Mark Zuckerberg are all doing today. They are blocking off weekly time to read books, learn new technology, experiment and get this, just to sit and think.

We are 3.0-ing everything in sight, from education, to healthcare, to finance, to travel, to manufacturing, to entertainment with rapidly developing technology most people barely fully understand. Experimentation and trial and error may have to start becoming a fundamental business strategy.

Google got it right when they set aside a day a week called the 20% time, where employees indulge in free will, passion and experimentation. It is both surprising and obvious that 50% of Google's products (Gmail, Adsense, Google Maps, Google Talk, and more) all started on that one day where they gave engineers room to play, solve problems and delight themselves.

Unfortunately, current reports claim that fewer and fewer employees are taking advantage of 20% time at Google, because of career driven managers, productivity stats and the oh so powerful "pressing present" which includes breakneck product shipping and career moves which prioritize short term gains over long term gains.

Here again, lies yet another paradox. We all know the biggest, richest fish in the pond no longer corner the market. As systems evolve, nimble startups with no legacy software to maintain, no deathly employee ennui, and no preconception of how problems should be solved are part of the level playing field that seemed impossible just a short time ago. One of the most popular games in the world is Minecraft, a game programmed single handedly by Markus Persson in his basement which later sold for $2.5 billion to Microsoft.

The Goliath problem in big companies has gotten so threatening, that consulting firms like Bionic Solution are being hired by Fortune 500 companies to start rogue startups within big companies. They often start by cherry picking a small multidisciplinary group of free thinking employees, form a B Corp company around them, and then lock them in a room together. They remove rules, pressure to succeed, and preconceived product plans from above, and let creativity and synergy do it's work. Once the teams emerge with tested prototypes, they have a massive company behind them to scale things up.

Multidisciplinary Teams Win At the time Fink started publishing the first paper edition of this book in 2018, the embarrassing reality was that the creators and many people exploring the technology weren't really sure what to do with AR quite yet.

Augmented reality is nothing more than a modern pencil. A substrate. A platform similar in many ways to the soon to be outdated Internet. So now what?

The "why" behind the technology is worth thinking through. In the digital age, exponential technology has the ability to replicate and magnify systems at speeds and scales we don't fully grasp. So what needs do we have today? What areas do we want to advance? What do we want to say?

Definitively, the main fear we should have about augmented reality is that as we wake up into our future tomorrow of big possibilities, that we shoot tragically low. How do we use augmented reality to redefine not education, but our wildest dreams for education?

Can augmented reality do more than offload daily drudgery? Bring delight to our friends? Or give real time superhuman insights to everyday people? Can we use AR to overcome our fears, break bad habits, travel through time and embed higher values into products themselves? Yes. Can we use it to cure or prevent terminal illnesses, spark a consciousness revolution, save planet earth, or make meaningful art together in groups? Of course.

Modern innovation happens when you get a bunch of passionate experts from varied fields who are trying to solve important problems together with experts in augmented reality and other frontier technologies. The why behind the technology becomes clear, not to mention combining augmented reality with computer vision, chatbots, blockchain, 3D printing, artificial intelligence, quantum computing and crowdfunding, producing even more amazing realities.

So what does augmented reality have to do with truck drivers losing their jobs? Everything. All this frontier tech is interconnected, and so are we. How do we solve problems together? How do we become our own government? How do we

balance ourselves? How do we all work less? How do we share? Can we imagine new rules, new frontiers, something better? Can computers help us?

The expression "think outside the box" is now also broken down on the side of the road. We can finally start to "think outside of the system" of default possibilities. If we create time to stop to consider our vision statement for the future, and what that looks like in detail, we actually have a good chance of making it real.

Proceed into the future, curious, and undaunted. Partner your human intelligence and wisdom with the smartest kid in the room (our computers). Grab this book, and the metaverse you have been invited into and jump-in-where-you-can, do-what-you-can, learn-what-you-can into your very own success narrative. Dream it, plan it, do it.

With passion, purpose and augmented reality, we can truly redefine reality and learn to alchemize beyond the frustrating paradigms of the past.

After all, the only way of discovering the limits of the possible is to venture a little way past them into the impossible.. - Arthur C. Clarke

See you on the other side (of Fink's unfolding metaverse).

Zenka

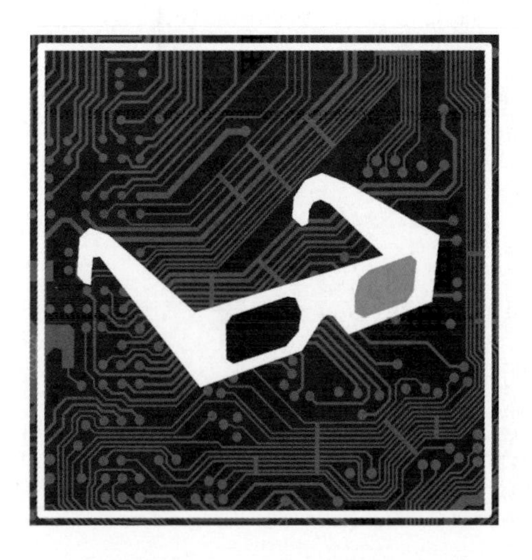

scan
this

APPENDIX 1

My augmented reality work has gone from the streets, to indoor installations, to schools, to gallery walls, and back around again. It all began in 2013 when I started cementing augmented reality ceramic murals in the streets of Los Angeles, back when people were still confusing augmented reality with QR codes. Over the years, I have loved watching people delight in the fact that AR can so gracefully edit our world. Creating AR art is like reinventing physics, obliterating scarcity, and unlocking our imagination. See you in the new reality. Z

Zenka's AR lino prints can be purchased online at: www.zenka.org/artshop

To see augmented art in the following appendix, with the app open, hold your device over the artist's design.

Sleep
Later

P + ½ρV² = consta
lift = C_L × (½ρV²) × S
drag = C_D × (½ρV²) × A
NH
C=N=N-C
NH
NH
Growth

A/P
"Hololens Hackathon"
2017

A/P
"Coco and the River Dragon"
2010

11/40
"Leaping into Magic"
2016

1/2
"Seguir"
2017

Water Lily Invasion (Augmented Reality 2013 - 2018, Tamiko Thiel)
tamikothiel.com/AR/waterlily.html

As global water levels and temperatures rise, plants and animals are mutating to adapt. Strange new creatures are arising at the interstices between plant and animal, questioning and transgressing the boundaries of what is considered to be reactive flora or active fauna. First seen in 2013 on the Star Ferry in Hong Kong, the seemingly small, innocuous, but extremely invasive water lily is apparently sensitive to the mediated human gaze: It can enlarge to engulf viewers who look at it in the display of their mobile app. It is hypothesized that it feeds off electromagnetic energy from mobile devices. Until now, viewers have reported feeling no effects other than a temporary fibrillation. Scientists worry, however, that future mutations could cross boundaries between insentient plant and sentient, perhaps even carnivorous life forms. This is the first instance captured in book form.

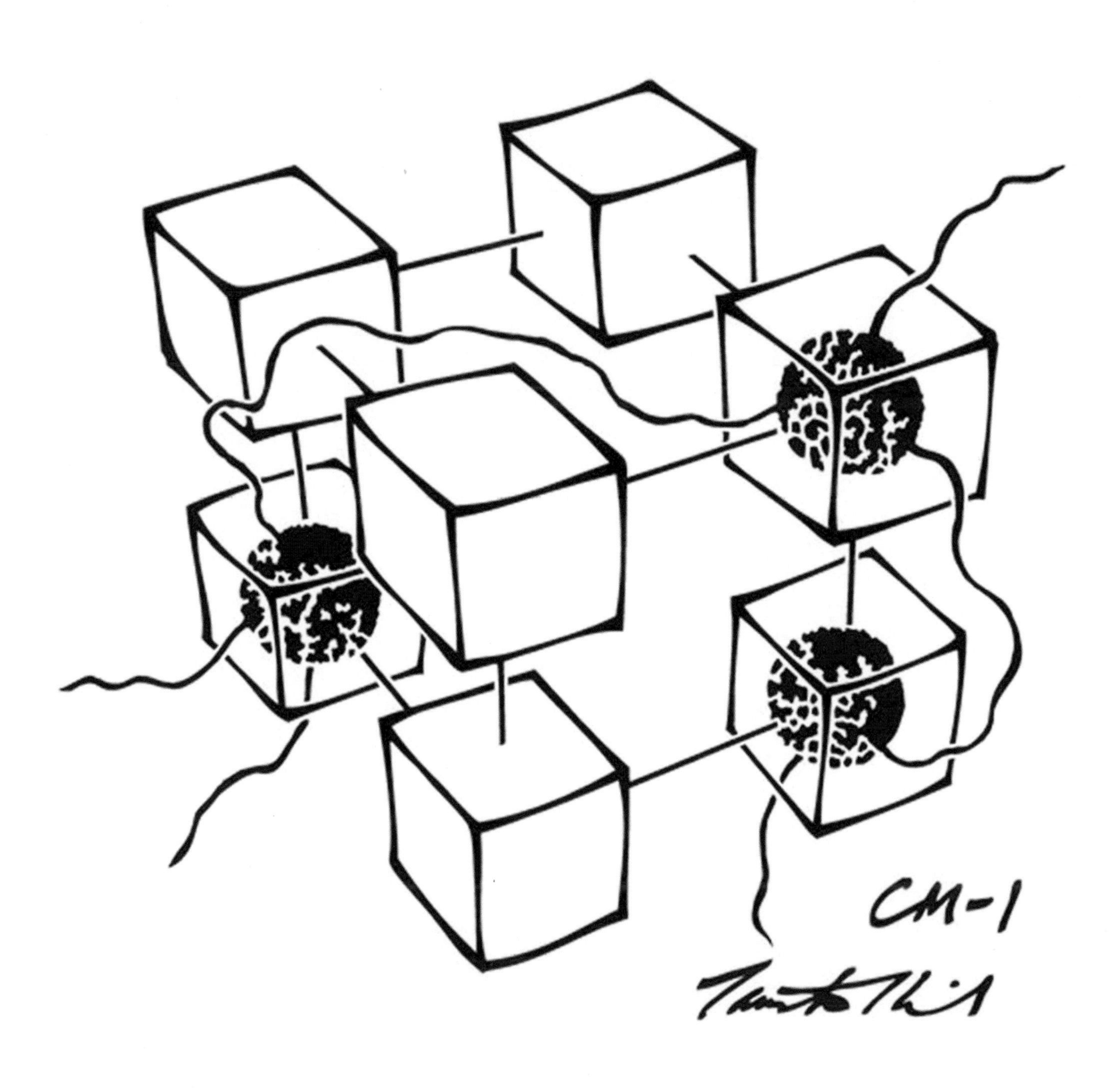
CM-1

Connection Machine CM-1/CM-2 (Machine 1986/1987 Thinking Machines Corp; Augmented Reality 2018 Tamiko Thiel)
tamikothiel.com/cm/

The Connection Machine CM-1/CM-2 was the only supercomputer whose form was designed after a t-shirt. It was invented by Danny Hillis as the first commercial supercomputer created to worsk on problems of artificial intelligence, and built at his startup Thinking Machines Corporation. Its computer architecture was inspired by the human brain, with 65,536 tiny processors connected in an internal 12-dimensional boolean hypercube network designed by Nobel physicist Richard Feynman. Lead product designer Tamiko Thiel created the t-shirt logo (called the "Feynman t-shirt" after Apple used a photo of him wearing it in their 1990s "Think Different" campaign) to reflect how the software data connections were independent of the hard-wired cube-of-cubes network structure. When it came time to build the machine itself, Thiel and her team echoed the t-shirt design in the actual form of the machine: A matt black cube-of-cubes, with translucent doors through which you could see the machine "thinking" as the 4,096 internal processor chip status lights blinked due to processor activity. CMs are preserved in the permanent collections of the Museum of Modern Art New York, the Computer Museum in Silicon Valley and the National Museum of American History/ Smithsonian Institution. Hardware and software concepts from the Connection Machine went on to substantially influence big data and artificial intelligence companies like Google, cluster and cloud computing, and specialty integrated circuit chips for graphics processing and artificial intelligence applications.

Charlie Fink

Charlie Fink writes about VR, AR and new media for Forbes. He is the former Disney, AOL, and AG Interactive executive who famously came up with the idea for The Lion King, "Bambi in Africa". In the 1990s, Fink was EVP & COO of LBVR pioneer Virtual World Entertainment. He is now a consultant and frequent speaker on the topic of VR & AR at conferences like CES, SXSW, AWE, ARinAction, On the Lot (keynote), VR Toronto (keynote), VR for Good (keynote) and VR Voice Entertainment Summit (keynote, co-organizer).

In addition to his column in Forbes, Fink is the contributing editor for The VR Voice, and a regular contributor to HuffPost, Virtual Reality Pop, and VR Focus. He is the author of Charlie Fink's Metaverse, an AR Enabled Guide to VR & AR, continuously updated at FinkMetaverse.com

charliefink.com
@charliefink

Robert Fine

In 2011, Robert launched the only printed magazine covering social media, The Social Media Monthly. In January 2014, he launched his second print title, The Startup Monthly. In May 2016, he launched VRVoice.co, a content vertical on all things virtual reality. In addition to his publishing endeavors, Robert continues to provide IT strategic planning consulting services to both the private sector and non-profit communities.

Robert was the Senior Director of Global Strategy and Development of IT at Conservation International (CI). Robert joined CI in 2000 to take responsibility for connecting all of CI's forty field offices to the Internet with broadband connectivity. During his 10 year tenure, he built an international staff of 25 IT professionals.

Robert has over ten years of additional work experience as a systems and sales engineer with various companies including CMGI, Hughes Network Systems, ioWave and Raytheon, as well as two of his own consulting companies: GeoPlan, Inc., and The Cool Blue Company. Robert has a bachelor's degree in mechanical engineering from Villanova University, a master's degree in environmental science and public policy from Johns Hopkins University, and is ABD at George Mason University.

linkedin.com/in/bobfine
@bobfine

Philip Rosedale

Philip Rosedale is CEO and co-founder of High Fidelity, Inc., a company devoted to exploring the future of next generation shared virtual reality. Prior to High Fidelity, Rosedale created the virtual civilization Second Life, populated by one million active users generating US$700 million in annual transaction volumes. In addition to numerous technology inventions (including the video conferencing product called FreeVue, acquired by RealNetworks in 1996 where Rosedale later served as CTO), Rosedale has also worked on experiments in distributed work and computing.

linkedin.com/in/philiprosedale
@philiprosedale

Ontario Britton

Ontario Britton develops apps that are meaningful and tests the boundaries of new technologies like AR. As a father of six boys, he especially likes developing software that they can learn from and enjoy. He helped launch more than twenty iOS apps in Swift, C++, C# and Obj-C with millions of downloads among them. Apps he has contributed to have garnered some of the industry's highest recognition and have been featured on the App Store numerous times. One educational children's app, Endless Alphabet, was awarded Apple's coveted "App Store Best" of 2013. He is most fulfilled by creating novel 3D experiences, especially those in mixed reality.

linkedin.com/in/ontariobritton
ontariobritten.com

Matt Miesnieks

Matt Miesnieks is a serial AR entrepreneur and investor in early stage emerging technology startups. He is CEO of 6D.ai, and a partner at SuperVentures, investing only in AR startups. 6D.ai is a stealth AR company that lets applications interact with the real world, working in partnership with Oxford University's advanced computer vision research. He led an AR R&D team at Samsung and previously co-founded Dekko, the company that first invented 3D AR on the iPhone. Prior to this, Matt led worldwide customer development at Layar in Amsterdam, resulting in their AR browser being pre-loaded on the majority of Android smartphones. Before that Matt was a VP & GM in Asia leading commercial and technical teams in the smartphone industry.

linkedin.com/in/mattmiesnieks
@mattmiesnieks

Stephanie Llamas

Stephanie Llamas is Vice President of Research and Strategy and head of immersive technology insights at SuperData Research. As VP, Stephanie oversees research and analytics across all digital gaming and interactive media segments for the company. She is a digital industry veteran, working in social media and digital art prior to joining SuperData. With more than 10 years of professional experience in digital media and a background in scholarly research, she brings a unique combination of academic acumen and business savvy to the industry.

Stephanie is one of the first analysts to provide market intelligence on the current generation of immersive technology, leading SuperData to become the gold standard in XR research. A thought leader and speaker, Stephanie regularly shares her professional insights with leading outlets such as Forbes, The Wall Street Journal, and The New York Times, and at major industry events such as CES and SXSW.

Stephanie received her B.A. in Art Theory and Practice from Northwestern University and her M.A. in Media, Culture and Communication at NYU. She lives in Brooklyn, New York with her husband Carlos and pitbull Yuna.

linkedin.com/in/stephaniellamas
@staphinaners

Michael Eichenseer

Michael Eichenseer writes about VR and helps startups tell their stories. With an education in information technology, Michael has both founded and worked with companies in the nutrition and fitness, legal cannabis, and chatbot industries. A lifelong gamer gone fitness enthusiast, he believes VR and AR can repair and improve the relationship between technology and the human body. Michael focuses his work at the intersection of game design and human behavior.

linkedin.com/in/mikenseer
@mikenseer

Dr. Annika Steiber

Dr. Annika Steiber has over 18 years of executive management experience from both startups and larger companies in the private high-tech sector. She is currently the Founder and CEO of the company A.S. Management Insights, a company with the mission to develop thought leadership in the areas of management for the digital economy and innovation management. Dr. Steiber is also a Managing Director at Berkeley Research Group, a thought leader founded by one of the world's most cited management researcher professors, David Teece.

Dr. Steiber has a PhD from Chalmers University of Technology in Sweden. She is an international authority in the field of management and innovation, lecturer at Santa Clara University in the USA, and the author of several award-winning research articles and the two well known management books: The Google Model (Springer 2014 and Vinnova 2014), The Silicon Valley Model-Management for Entrepreneurship (Springer 2016 and Liber 2016) and Management for the Digital Age: Will China surpass Silicon Valley?.

www.annikasteiber.se

Mark Billinghurst

Mark Billinghurst is Professor of Human Computer Interaction at the University of South Australia in Adelaide, Australia. He earned a PhD in 2002 from the University of Washington and researches innovative computer interfaces that explore how virtual and real worlds can be merged, publishing over 350 papers in topics such as wearable computing, augmented reality and mobile interfaces. Prior to joining the University of South Australia he was Director of the HIT Lab NZ at the University of Canterbury and he has previously worked at British Telecom, Nokia, Google and the MIT Media Laboratory. His MagicBook AR project, was winner of the 2001 Discover award for best entertainment application, and he received the 2013 IEEE VR Technical Achievement Award for contributions to research and commercialization in augmented reality. In 2013 he was selected as a Fellow of the Royal Society of New Zealand.

Ori Inbar

Ori Inbar is founder of the first early stage fund dedicated to augmented reality (AR).
Ori is also the co-founder and CEO of Augmented Reality.ORG, a global non-for-profit organization dedicated to advancing augmented reality (AR).
Augmented Reality.ORG's mission is to educate and promote the true potential of AR, and hatch augmented reality startups that offer unique value to its active users.
In 2009, Ori was the co-founder and CEO of Ogmento, one of the first venture-backed companies conceived from the ground up to develop and publish augmented reality games: Games that are played in the real world.
Ori has been an enterprising champion of the augmented reality industry since 2007. He established Games Alfresco, a leading augmented reality blog that helped popularize AR, and in 2010 co-founded the Augmented World Expo (Formerly ARE), the world's largest and most influential conference for AR, now in its 8th year.
Ori is a recognized speaker in the AR industry, lecturer at NYU, as well as a sought after adviser and board member for augmented reality startups.

linkedin.com/in/oriinbar
@comogard

Peter Wilkins

Peter is an experienced technology leader with over a decade of software engineering and management experience spanning feature animation, AAA video games, and VR/AR. He has led product development teams to successfully execute complex technical roadmaps with high degrees of risk and uncertainty. A strong cross functional collaborator, he often sits at the intersection of engineering, design, art, and business. He is credited on Academy Award nominated films and helped invent the 360 video genre in 2013 by creating the first 360 3D film, featured at Sundance New Horizons.

emergentvr.com
@peterawilkins

Dirk Schart

Dirk Schart is Head of PR & Marketing at RE'FLEKT, the Munich-based tech company building the augmented and mixed reality enterprise ecosystem. Dirk is co-author of the book Augmented and Mixed Reality as well as a recognized speaker at events such as Audi MQ, Augmented World Expo, DLD Conference, dmexco and Lufthansa's Flying Lab to SXSW, where he gave a keynote aboard an Airbus A380 and showed augmented reality live at 11,000 feet. Dirk's vision is to provide humans with unlimited access to knowledge through augmented, mixed and virtual reality and to create a world where we naturally interact with the new realities.

linkedin.com/in/schartdirk
@dirkschart

Samuel Steinberger

Samuel Steinberger is a producer, editor, and journalist whose video and writing bridges the intersections between technology, business and travel. A graduate of Columbia University's School of Journalism, he is currently an associate producer for Soledad O'Brien, where he works on both short and feature-length documentaries. He is focused on using new storytelling mediums as virtual and augmented reality continue to shape the media landscape. When not writing or filming, Samuel hides from screens in the great outdoors. He is based in New York City.

samuelsteinberger.com
@slsteinberger

Dr. Walter Greenleaf

Walter Greenleaf, PhD is a research neuroscientist and medical product developer working at Stanford University. He is known internationally as an early pioneer in digital medicine and virtual environment technology. With over three decades of research and product development experience in the field of medical virtual reality technology, Walter is considered a leading authority.

Dr. Greenleaf is currently a Distinguished Visiting Scholar at Stanford University's MediaX Program, a Visiting Scholar at Stanford University's Virtual Human Interaction Lab, Director of Technology Strategy at the University of Colorado National Mental Health Innovation Center, and Member of the Board of Directors for Brainstorm: The Stanford Laboratory for Brain Health Innovation and Entrepreneurship. Previously, Walter served as the Director for the Mind Division, Stanford Center on Longevity, where his focus was on advancing research on age related changes in cognition.

In addition to his research at Stanford University, Walter is Senior Vice President of Strategic & Corporate Affairs to MindMaze and Chief Science Advisor to Pear Therapeutics. He is a VR technology and neuroscience advisor to several early-stage medical product companies, and a co-founder of Cognitive Leap.

about.me/waltergreenleaf
@waltergreenleaf

Tim Kashani

As founder of IT Mentors and of Apples and Oranges Studios/Arts with Broadway actress Pamela Winslow Kashani, Tim creates work environments where people with diverse artistic and technical skills unite to create innovative and entertaining products in business and the arts. With IT Mentors, his global technology training company, Tim travels all over the world training and designing systems for some of the world's largest technical and financial corporations. With Apples and Oranges, Tim produced the Tony Award winning Broadway productions of Hair, Memphis and An American in Paris.

Tim treats artists as entrepreneurs, empowering them with the resources to succeed. Tim and his team created the Theatre Accelerator, seeking to apply the methods, processes, and mindset of the startup world to the performing arts. This includes data driven ticket sales, marketing, social media, and most importantly, a call to rethink the entire developmental and distribution cycle of new stories and explorations in how AR/VR/MR expand past the fourth wall.

linkedin.com/in/timkashani
@timkashani

Tony Parisi

Tony Parisi is a virtual reality pioneer, serial entrepreneur and angel investor. Tony is the co-creator of 3D graphics standards, including VRML, X3D and glTF, the new file format standard for 3D web and mobile applications.Tony is also the author of O'Reilly Media's books on Virtual Reality and WebGL: Learning Virtual Reality (2015), Programming 3D Applications in HTML5 and WebGL (2014), and WebGL Up and Running (2012).
Tony is currently Head of VR and AR at Unity Technologies, where he oversees the company's strategy for virtual and augmented reality.

tonyparisi.com
@auradeluxe

Kyle Melnick

Kyle Melnick graduated from St. John's University in NYC and is currently a writer for VRScout.com covering interactive new media such as VR, AR and MR technology. This includes some of the most exciting recent breakthroughs in immersive technology and gaming. His work extends beyond writing, acting as a producer, editor and on-air host for gaming and technology centered news and entertainment. In addition to his involvement in technology, he's also a working actor with credits including the The Late Show, Mohegan Sun, Hasbro and Nickelodeon.

@klyeonfire

Samantha G. Wolfe

Samantha G. Wolfe is CEO of PitchFWD Reality, a VR/AR consulting agency. Sam is a creative strategist driven to empower companies and executives to make the never-been-done-before a reality. She identifies and integrates innovative tools and trends, strengthens brand positioning, establishes marketing strategies, and secures revenue generating partnerships. Sam recently co-authored the new book, Marketing New Realities: An Introduction to Virtual Reality & Augmented Reality Marketing, Branding, & Communications. Throughout her career, she has managed B2C and B2B campaigns for Showtime Networks, Food Network, Cooking Channel, TV Guide, Rovi (now TiVo), and RLTV.

linkedin.com/in/samanthawolfe
@samanthagwolf

Tim Merel

Tim Merel is an AR/VR, mobile and games software engineer, investment banker and entrepreneur, with education in computer science, law and business from Yale and Sydney University. Tim is a global super connector, with C-level relationships across startups, corporates and VC/private equity. Tim leverages his knowledge, relationships and ideas across America, Asia and Europe for industry reports, strategy consulting and investment banking and has worked on over 60 deals globally. Tim is a global AR/VR, mobile and games expert, equally comfortable at discussing strategy, operations, technology and deal structures. As well as those companies he's founded, Tim previously worked for News Corporation and Ernst & Young.

digicapital.com
@digi-capitalist

Tamiko Thiel

Tamiko Thiel is an internationally acknowledged pioneer creating poetic spaces of memory for exploring social and cultural issues with new media. Her first virtual reality work was Starbright World (1994-97, with Steven Spielberg). Her VR installation Beyond Manzanar (2000, with Zara Houshmand) is in the permanent collection of the San Jose Museum of Art in Silicon Valley. She has worked with augmented reality since 2010, first creating ARt Critic Face Matrix for the path-breaking AR intervention We AR in MoMA at MoMA NY, and then in countless commissions and interventions worldwide. In 2017 she is Google VR Tilt Brush Artist in Residence and her Connection Machine CM-2 supercomputer is exhibited as part of the permanent collection of MoMA NY.

tamikothiel.com
@tamikothiel

Zenka

Zenka is an artist and social architect. She is a thought leader in the future of augmented reality technology and exponential change. Her focus is creating meaningful conversations around what is possible in the future using emerging technology and crowd participation. Her work can be seen in the permanent collection of Delta at JFK, Accenture Interactive in Soho, The Museum of Tech and Innovation, and tech startups in the US and Europe. She has spoken at TEDx and tech conferences around the country about the importance of leveraging our unique place in time.

zenka.org
@hellozenka

Original AR Animation by Living Popups

Producers
Jamie Dixon
Cheryl Bayer

Art
Michael Scharf

Animation
Ken Pellegrino

Writers
Hunter Maats
Katie O'Brien

Voices
Wade: Alan Weischedel
Ripley: Susie McDonnell
Charlie: Charlie Fink

Production and Business Development
Paula Davidson
Peter Coleman
Alice Frankston
Katie Rowbotham

Software Developer
Christian Aubert

Living Popups is an interactive augmented reality (AR) content platform and media company. Living Popups specializes in being a premier storyteller, blending content with interactive and enhanced features. Living Popups is a partnership of the best and brightest creative content providers in Hollywood with thoughtful and innovative educators. Together, we make compelling storytelling that comes to life.

Cheryl Bayer – CEO

Cheryl Bayer is an entertainment industry leader in casting and developing talent, marketing, and branding campaigns. She started with her own casting company and then moved on to the Head of Talent and Development at ABC Productions, packaging agent at CAA, and Senior VP of Comedy at Fox. She was also pivotal in launching ESPN W and produced an award winning civil rights documentary, Bookers Place. She was a key player in bringing us *In Living Color, Baywatch, Dream On, Roseanne, My So Called Life, Home Improvement, Malcolm in the Middle, That 70's Show* and *Family Guy*. Her current projects include developing shows for MTV, and Michael Eisner's Tornante.

Jamie Dixon – CTO

Jamie Dixon, a member of the Motion Picture Academy, is a leading innovator in using cutting edge technology in the field of entertainment. From innovating 3D stereo computer graphics and animation to creating digital scanning, printing and production techniques that were among the earliest stirrings of modern digital filmmaking, he is always exploring what technology can add to the storytelling process. Currently, his application of character based storytelling to AR is opening up new avenues of socially immersive entertainment. His visual effects work can be seen in on over 100 major motion pictures. Jamie's creation of the face transitions in Michael Jackson's *Black or White* music video made the word "morphing" a permanent part of our cultural lexicon.

Peter D. Coleman – Head of Production

Peter D. Coleman is a content creator, executive and producer of television, digital media and independent feature films. Peter co-founded and was COO of What The Funny, a digital comedy content company along with Marlon Wayans and Funny or Die co-creator Randy Adams. He led the company from inception to over one million monthly views and oversaw production of more than 350 pieces of content. Peter served as a Press Advance Person for the Clinton White House, working directly with the President and First Lady, and multiple foreign governments and NGOs. Peter began his career as a Network Page at NBC for David Letterman, SNL and Nightly News. He has a Masters Degree in Journalism from Columbia University.

Katie O'Brien – Creative Content Officer

Katie is a Harvard graduate, magna cum laude no less, who has combined her love for creating content and working with high school students. Katie soon realized no one was concerned with showing kids how to learn based on what works for them, individually. So she and Hunter co-authored a book dedicated to just that, The Straight-A Conspiracy: Your Secret Guide to Ending the Stress of School and Totally Ruling the World. She has also followed her passions of writing, directing, and performing, and has co-created a number of series for The Hub, IFC, ABC Digital, and the web.

Hunter Maats – Creative Content Officer

Hunter is a Harvard educated writer, podcaster, entrepreneur and all over witty raconteur. He researched cancer with James Watson, co-discoverer of the DNA double helix, discovering that the divide in communication between scientists and the general public stands in the way of progress. This is a topic that became the focus of his career. He founded Overqualified Tutoring, which applied the latest findings in neuroscience and psychology to help students of any age succeed. He currently co-hosts with comedian Bryan Callen the podcast, *Mixed Mental Arts*, reaching 250,000 downloads a month.

Ken Pellegrino – Animation Development

Ken began his animation career as a modeler for a small boutique VFX company based in Los Angeles. His talents soon led him to being made Lead Artist for Motorola's new product release packages and he worked on projects for Absolut Vodka, T-Mobile, HP, and Disney theme park commercials. He found a creative home at Hammerhead Productions as a CG supervisor and spearheaded the effort to use new technologies to simplify animation using advanced GPU rendering technology.

Christian Aubert – App Development

Christian is an all around technological artist with a 20-year history in visual effects, computer graphics and associated software development. He set up the virtual production stage for *Avatar*, developed a complete asset and project production management system for Origami Digital, built rendering production pipelines for Sony Pictures Imageworks, Henson Studios, and many more. He is a skilled VR/AR artist, has automated the workflow and built ride-film applications targeted to Oculus and Gear VR for Threshold Entertainment and Pure Imagination Studios. He has also implemented real time hair/fur on PlayStation 3. He is also lead developer of Argonaut and Bubo products for gait analysis.

3 Degrees of Freedom (3DOF) Three degrees of freedom represents the ability of an object's rotation, pitch, and yaw to be tracked, usually with a gyroscope.

360 Video Animations or videos that film 360 degrees of content, completely surrounding a viewer. Viewers generally can only view this content from a stationary position and cannot interact with it, although content that uses volumetric capture can facilitate some levels of interaction.

5G The 5th generation of mobile networks is a proposed telecommunications standard setting mobile data rates from tens of megabits per second up to 1 gigabit per second.

6 Degrees of Freedom (6DOF) Hardware and content that tracks a user's head and body position. This includes directional positioning (front, back, left, right, up and down) as well as rotation (pitch, yaw and roll).

Artificial Intelligence (AI) The theory and development of computer systems able to perform tasks that normally require human intelligence, such as visual perception, speech recognition, decision making, and translation between languages.

Active Matrix Organic Light Emitting Diode (AMOLED) An active matrix organic light emitting diode is a more energy efficient OLED with faster refresh rates useful for virtual reality displays.

AR Cloud The collection of proposed technologies to enable seamless spatial computing via augmented and mixed reality displays throughout the physical world.

ARCore Google's software development kit using the smartphone camera to provide motion tracking and depth sensing data to app developers.

ARKit Apple's software development kit using the smartphone camera to provide motion tracking and depth sensing data to app developers.

Augmented Reality (AR) Overlaying or mixing simulated digital imagery with the real world as seen through a camera and on a screen. Graphics can interact with real surroundings (often controlled by users).

Augmented Virtuality (AV) Incorporating elements from actual reality into a virtual environment.

Blockchain A digital ledger in which transactions made in bitcoin or another cryptocurrency are recorded chronologically and publicly.

Capacitive Sensor A sensor that measures if something is conductive. Used by device manufacturers to detect the presence and position of human fingers.

Cave Automatic Virtual Environment (CAVE) A virtual reality system that uses projectors facing the walls of a room surrounding a single user. The image projected changes based on the orientation of the user as they turn and move around the room.

Depth of Field Depicts the size of in-focus area between two distances in an image.

Depth Perception The ability to see the world in three dimensions to determine both near and far.

Escape Room A venue consisting of one or more rooms filled with clues, puzzles, and story elements designed to challenge players under a time constraint as they attempt to escape.

Eye Tracking Tracking a user's specific point of gaze, or whatever the user can see clearly and is not part of their peripheral vision.

Field of View (FOV) The physical range, measured in geometric degrees, within which a person can view the reality in front of them. In the real world, an average human has a horizontal FOV of 200 degrees, whereas the average high-end XR headset has a FOV between 70 and 110 degrees.

Foveated Rendering A developer technique that reduces a hardware's processing power by fully rendering only what a user is looking directly at, leaving the surrounding areas blurry (imitating peripheral vision). This technique requires eye tracking capabilities to precisely detect where the user is looking.

Frames Per Second (FPS) The speed at which video is displayed on a screen or through optics. Most VR applications are targeted at 90 FPS or more.

Free-to-Play The pricing model for games that allows access to players for free. Monetization of users can come from advertising, microtransactions, user data, etc.

Free Roam or Warehouse Scale VR Like room scale VR, Free Roam, or Warehouse Scale VR, gives the user the ability to be more present in a simulation by moving freely in a digital world. Like all high-end VR, the user is tethered to the PC. However, in warehouse scalable the PC is worn on the back. Users strap on markers, and a multitude of trackers are arrayed about a room roughly 10'x30'. In this way, the user may roam freely under natural locomotion, typically in a multiplayer environment.

General Public License (GPL) An open source software license allowing the free use, reuse, and modification of software.

Haptic Refers to the sense of touch simulated by technology.

Haptic Gloves Gloves which simulate physical touch with virtual objects and environments.

Head Mounted Display (HMD) A headset that covers a user's eyes and displays XR.

Hologram A hologram is a photographic recording of a light field, rather than of an image formed by a lens, used to display a fully three dimensional image of the holographed subject, which is seen without the aid of special glasses or other intermediate optics.

Head Related Transfer Functions (HRTF) The mathematical functions expressing the ability of the human ear to localize sound. Used to simulate three dimensional sounds using stereo interfaces.

Immersion Being absorbed in an experience such that awareness of reality is ignored or forgotten.

Inertial Measurement Unit (IMU) A sensor measuring the force and angular rate of an object.

Inside Out Facing VR and AR HMDs use outward facing cameras to first, in the case of VR, sense objects and other real obstacles of which a user needs to be aware. In the Microsoft Windows MR system, the camera tracks hand movements. Using the HoloLens, outward facing cameras not only detect gestures but allow remote experts to see what you are seeing in real time.

Latency The measurement of time between user input and computer output. Usually represented in milliseconds(ms). HMD optics requires a latency between the head movement and the visual display of 7ms or less to reduce motion sickness in users.

LiDAR Light detection and ranging measures distance via the time it takes light to bounce off an object.

Location Based VR (LBVR) VR experiences built for a dedicated space outside the home, such as theme park attractions, arcades and entertainment centers that can accommodate multiple experiences at once (i.e., IMAX VR). Refers to a family of attractions also called Location Based Entertainment (although this can be anything that's not a store), FEC (Family Entertainment Center) and Vrcades.

Locomotion The movement of an entity through space. In regard to VR users: The methods used in a given VR experience allowing a user to traverse the environment.

Massive Multiplayer Online Game (MMOG) A multiplayer game allowing a large population of players, hundreds or thousands, to play together in the same space.

Massive Multiplayer Online Role Playing Game (MMORPG) A massive multiplayer game consisting of role playing elements such as levels, player classes, guilds, quests, etc.

Metaverse The sum of all parts of the Internet that can be navigated by anyone with a proper VR rig. The word "metaverse" combines the prefix "meta" (meaning "beyond") with "universe", here referring to an infinite number of interconnected virtual or digital spaces.

Milgram Scale (Virtuality Continuum) A scale developed by Paul Milgram that explains the range between real environments and virtual environments (or what Milgram called mixed reality).

Mixed Reality (MR) An experience that combines a viewer's real environment with augmented reality or augmented virtuality using a headset.

Mobile AR AR rendered through a smartphone's camera and seen on its screen, also called handheld AR. Can also be done with a digital tablet.

Mobile VR Created specifically for mobile computing, mobile VR is the most popular form of consumer VR, examples include Google Cardboard and Samsung Gear VR.

Neural Interface Also known as a brain computer interface (BCI), refers to the ability of a user to input to a computer directly via their mind.

Neural Lace Proposed by Elon Musk's company Neuralink, is a mesh that exists on the brain providing a direct interface between the brain and technology.

OASIS Ontologically Anthropocentric Sensory Immersive Simulation is the fictional virtual reality depicted in Ernest Cline's novel Ready Player One that connects all other virtual experiences into a cohesive ecosystem.

Organic Light Emitting Diode (OLED) An organic light emitting diode is a display technology utilizing organic compounds which produce light when exposed to electrical current.

Olfactory Refers to the systems that make up the senses of taste and smell.

Open Source Software whose source code is openly available and free to use.

Optics Refers to the study and application of the properties and behavior of light. Encompassing the hardware that provides visuals in a HMD.

Outside In Tracking The tracking of position in virtual reality using sensors external from the device that tracks the position of the device in a physical space.

Peripheral Vision What the eyes see to the side of the center of vision (the fovea).

Photogrammetry Using the data from multiple high resolution digital camera passes, to create a three dimensional digital reconstruction of captured space. Avatars can move freely within an environment created with photogrammetry.

Photorealism Virtual imagery made to look indistinguishable from real life.

Point Cloud A set of points in three dimensions consisting of X, Y, and Z coordinates.

Refresh Rate The frequency at which an image or data is updated, usually represented in hertz (Hz). A 120Hz display can update 120 times per second.

Resolution (visual display) The number of pixels that makes up a display. Represented as dimensions (i.e. 1920x1080).

Room Scale VR Tracking a user's position and movements within a space that is roughly the size of an average household room and reflecting them in a virtual or augmented world. This can be done using external sensors or through a headset itself using inside out tracking.

Screen Door Effect The effect of a low resolution display on virtual objects which appear to be viewed as if through a bug screen on a real world screen door.

Simultaneous Localization and Mapping (SLAM) The tracking of position within an environment by scanning and generating a map of said environment in real time.

Social VR Refers to virtual experiences designed for users to connect, communicate, collaborate, or compete with each other.

Software Developer Kit (SDK) A set of tools that allows software developers to build applications for specific hardware or digital infrastructure.

Spatial Computing Refers to all technologies involved in computation in three dimensional space. (point clouds, SLAM, virtual reality, mixed reality, etc.)

Spatial Light Modulator Devices used to modulate properties of light. Used in virtual reality optics to provide depth of field.

Stereopsis The perception of depth created by the overlapping of two visual feeds, such as two eyes spaced apart.

Stereoscope A device that displays a separate image to each eye to produce a three dimensional effect.

Tabletop AR AR (often activated by pointing a camera at a symbol such as a QR code) that is anchored onto a flat surface so the viewer can interact with it at all angles.

Telepresence Using telecommunications technologies to project one's self to another physical location digitally.

Tracking The determination of an object's position within three dimensional space.

Visual Inertial Odometry (VIO) The use of visual data to determine position in an environment.

Visual Positioning Service (VPS) Google's system to map environments in three dimensions. Used in their ARCore to position virtual objects in reality.

Virtual Reality (VR) An experience that requires a headset to completely replace a user's surrounding view with a simulated, immersive, and interactive virtual environment.

Volumetric Capture Also referred to as volumetric videogrammetry is the process of capturing a person volumetrically in 3D from an array of cameras pointing inward is called volumetric capture. Volumetric capture is video based, which makes the approach photorealistic and scalable. It allows us to reconstruct the person being recorded with full volume and depth, color and light, so that when viewed in a VR or AR headset, or in mobile AR, the person looks real, as if they are actually there, and you can walk around them and see them from any angle.

VR Sickness Also referred to as motion sickness, VR sickness is caused when the human vestibular system is thrown out of balance by the visuals a user sees not aligning with the movement they feel.

VRcade Refers to virtual reality arcades and LBVR installations.

X Reality (XR) A term that encompasses VR, AR, MR and everything in between. X is a stand-in for the unknown and, therefore, allows any pertinent variable to replace it. Traditionally, X Reality referred to bio-augmentation.

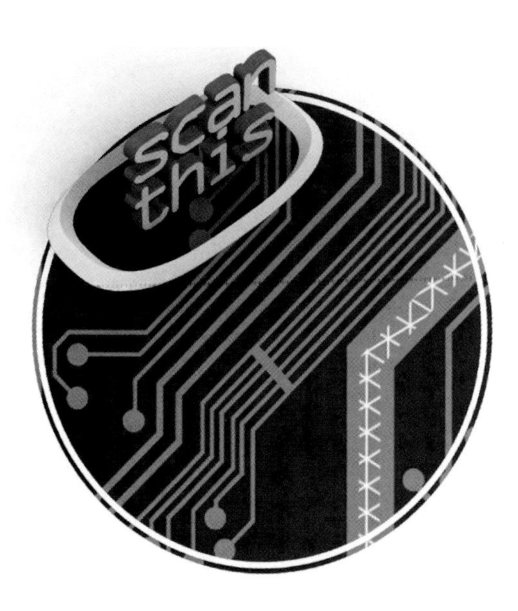

INDEX